Glycomimetics: Modern Synthetic Methodologies

ACS SYMPOSIUM SERIES **896**

Glycomimetics: Modern Synthetic Methodologies

René Roy, Editor
Universiy of Québec at Montréal

Sponsored by the
ACS Division of Carbohydrate Chemistry

American Chemical Society, Washington, DC

Library of Congress Cataloging-in-Publication Data

Glycomimetics : modern synthetic methodologies / René Roy, editor ; Sponsored by the ACS Division of Carbohydrate Chemistry.

p. cm.—(ACS symposium series ; 896)

Includes bibliographical references and index.

ISBN 0-8412-3880-4 (alk. paper)

1. Carbohydrates—Synthesis—Congresses. 2. Glycosides—Synthesis—Congresses. 3. Biomimetics—Congresses.

I. René Roy, 1952- II. American Chemical Society. Division of Carbohydrate Chemistry. III. Series.

QD322.S95G59 2004
547′.780459—dc22 2004057493

The paper used in this publication meets the minimum requirements of American National Standard for Information Sciences—Permanence of Paper for Printed Library Materials, ANSI Z39.48–1984.

Distributed by Oxford University Press

PRINTED IN THE UNITED STATES OF AMERICA

Foreword

The ACS Symposium Series was first published in 1974 to provide a mechanism for publishing symposia quickly in book form. The purpose of the series is to publish timely, comprehensive books developed from ACS sponsored symposia based on current scientific research. Occasionally, books are developed from symposia sponsored by other organizations when the topic is of keen interest to the chemistry audience.

Before agreeing to publish a book, the proposed table of contents is reviewed for appropriate and comprehensive coverage and for interest to the audience. Some papers may be excluded to better focus the book; others may be added to provide comprehensiveness. When appropriate, overview or introductory chapters are added. Drafts of chapters are peer-reviewed prior to final acceptance or rejection, and manuscripts are prepared in camera-ready format.

As a rule, only original research papers and original review papers are included in the volumes. Verbatim reproductions of previously published papers are not accepted.

ACS Books Department

Contents

Indexes

Preface

The recent successes encountered with carbohydrate vaccines, antimicrobial adhesion agents, and inhibitors of carbohydrate processing enzymes have paved the way for promising uses of carbohydrate mimetics as novel therapeutic agents. The increasing demand for metabolically stable glycomimetics, coupled with inclusions in the field of modern synthetic methodologies, together with the "reactivation" of "old" carbohydrate-containing antibiotics such as aminoglycosides and polyketide-derived aryl *C*-glycosides have triggered the interests of a new generation of "glycoscientists". By their critical positioning at the cell surfaces, carbohydrates are widely involved in several recognition phenomena responsible for biochemical processes working in good "faith". Paradoxically, mammalian cells express only a handful of carbohydrate residues. Thus, Nature has found a way to multiply the number of epitopes that could be composed out of these few sugars. Moreover, the number of tertiary structures and the valencies of exposed carbohydrates add to the complexity of potential recognition sites.

Therefore, the "playground" of carbohydrate chemists, initially taught to be rather bothering, happen to be far more complex than originally anticipated. This situation may explain the lag time necessary for the discovery of new carbohydrate-based drugs. The therapeutic game, as it stands today, as to consider several factors not previously encountered by traditional medicinal chemists. The name of the game is now that glycochemists have to take into account for the poor lipophilicity of carbohydrates, their tremendous conformational mobilities, and perhaps, equally important, the valencies of the carbohydrate epitopes necessary for selective and high-affinity ligand design. Equally important is the need for metabolically stable analogs, thus the importance of *C*-linked glycosides, conformationally restrained analogs, and multivalent structures.

This book is the result of combined symposia held under the auspices of the American Chemical Society (ACS). The symposia were held at the 225th National Meeting in New Orleans, Louisiana on March 23–27, 2003. The first symposium presented modern organometallic chemistry in the design of glycomimetics and their uses in asymmetric catalysis. M. H. D. Postema (Wayne State University) and R. Roy (Université du Québec à Montréal) presented their work on transition metal catalyzed reactions wherein ring closing metathesis (RCM) and palladium catalyzed cross-coupling reactions were discussed. A. Fürstner (Max-Planck-Institute, Germany) could not attend the Symposium but was able to join us at the 226th Meeting in New York for the M. L. Wolfrom Award Lecture given by me. The content of his lecture on olefin metathesis is included herein. The second symposium was the Hudson Award Symposium and the Award lecture delivered by R. J. Linhardt, now at the Rensselaer Polytechnic Institute (Troy, New York) was retained. The third symposium was co-organized by P. P. Deshpande (Bristol-Myers Squibb, New Jersey) and P. Seeberger (now at ETH, Zurich, Switzerland) and dealt with the "Chemistry and Applications of *C*-Glycosides". Unfortunately several participants could not contribute to this book but their lectures highlighted and reinforced the concerns expressed above.

The main topics of these symposia were collected in this book and the themes were regrouped under the umbrella of *Glycomimetics: Modern Synthetic Methodologies*. The authors were asked to include key experimental details dealing with modern synthetic procedures, most of which describe the use of delicate transition metal catalyzed reactions. Thus this book will be a valuable tool for well-trained traditional glycochemists and particularly for newcomers in the field of glycomimetics. The subject index at the end of the book will help the readers retrieve the desired informations.

This book does not pretend to compete with exhaustive review articles given by keynote experts in the field, several of whom participated in this avenue. Rather, it combines multiple facets of modern synthetic carbohydrate chemistry of values to those involved in glycomimetic research.

Thus, the first chapter, presented by Fürstner, describes the refined uses of olefin metathesis catalysts for the syntheses of natural cyclitols (conduritols A–F) that are known to play important roles as secondary messengers. The syntheses of marine lipidic lactone ascidiatrienolide, the phytotoxic herbarumin and pinolidoxin, and the complex resin glycosides tricolorin and woodrosin were elegantly prepared from sugar-derived building blocks. The key feature of this chapter was the observations that the usual *E/Z*-stereoisomeric mixtures resulting from the RCM process could be fine-tuned with the proper choices of the ring size together with the appropriate metal alkylidene complexes. The chemistry of olefin metathesis was further extended in Chapter 2 by Postema et al. who describe the elegant syntheses of various *C*-glycosides and saccharides. A combination of esterification-methylenation (Takai–Tebbe reactions)-RCM was engaged toward the preparation of several *C*-linked glycomimetics, that included a trisaccharide analog.

The third chapter by Linhardt and co-workers describes the difficult syntheses of *C*-linked saccharides derived from the family of ulosides Neu5Ac, KDO, and KDN using samarium iodide-mediated Barbier-type coupling. The syntheses of potential *C*-linked cancer related vaccine precursors are illustrated. The key feature here is the high-stereoselective synthesis of α-C-sialosides and other ulosides. Chapters 4 to 7 concentrate on different methodologies to access *C*-glycosides. Thus, the use of the powerful glycosyl phosphates is described by Seeberger and co-workers for the preparation of electron-rich aryl *C*-glycosides, obtained by an O- to C-Fries rearrangement. In this way, a straightforward synthesis of a natural bergenin analog was achieved. Chapter 5 by Parker illustrates a reverse strategy in which addition of lithiated glycals serving as nucleophiles to substituted quinones provided key precursors to the antifungal papulacandin and chaetiacandin *C*-phenolic glycosides. Key observations conducted to regioselective glycosylations of quinol silyl ethers toward phenol bis-glycosides. Chapter 6 by Franck and co-workers describes their pioneering work on the application of the Ramberg–Bäcklund rearrangement of glycosyl sulfones into *C*-glycosides by sulfur dioxide extrusion. The scope and limitations of the strategy are clearly discussed. Chapter 7 by Denton and Mootoo illustrates a combination of glycomimetic approaches to provide a new entry into the important family of antiinflammatory sialyl Lewis X analogs. They describe the synthesis of *C*-glycosides having restricted conformations, a key feature in the design of metabolically stable and high-affinity ligands. The tactic also relies on tethering the two saccharide partners by an ester linkage that is then transformed into an enol ether using Tebbe's titanocene complex. The ring closing is then effected by using a clever intramolecular hemithioacetal activation with Lewis acids. In this way, a 1,1-linked Gal-Man disaccharide was produced.

The last section of the book (Chapters 8–10) depicts slightly different methodologies to gain access to potent glycomimetics. Hence, Chapter 8 by Roy et al. elaborates on the transition metal catalyzed cross-coupling reactions to afford small glycoclusters related to the family of mannosides. Herein, three strategies are combined for improved ligand design. The first one relies on the improved affinity gained by subsite assisted binding of aryl mannosides coupled in the second by the scaffolding of multivalent saccharides onto conformationally rigid alkynylarene derivatives, the last one providing the third strategy. Preliminary biological data with the model phytoheamagglutinin Concanavalin A demonstrated the relative efficacy of the new "rigidified" glycoclusters. Chapter 9 by Bouvet and Ben goes along the *C*-glycoside chemistry applied to the synthesis of active *C*-linked antifreeze glycopeptides. Arrays of C-galactosyl peptides are compared to the natural glycoproteins. Finally, Chapter 10 represents an interesting novel idea in solid-phase glycopeptide syntheses wherein Guo describes the use of unprotected glycosyl amino acids as key building blocks and "phase tags". In this way, complex peptides, built by traditional solid-phase syntheses, are attached to free peptidoglycosides in *N*-methylpyrrolidinone (NMP) using standard coupling strategies. The resulting glycopeptides are then simply isolated by precipitation in ether.

Acknowledgements

We thank the following companies and associations for their generous contribution to the above Symposia: Strem Chemicals, Inc., Bristol-Myers Squibb Company, Princeton, New Jersey, Abbott Laboratories, Abbott Park, Illinois, and the ACS Division of Carbohydrate Chemistry. The contribution of anonymous peer reviewers is also particularly acknowledged.

René Roy
Department of Chemistry
Université du Québec à Montréal
P.O. Box 8888, Succ. Centre-ville
Montréal, Québec H3C 3P8, Canada
(514)–987–3000 ext. 2546 (telephone)
(514)–987–4054 (fax)
roy.rene@uqam.ca (email)

Chapter 1

Reflections upon Olefin Metathesis in Carbohydrate Chemistry

Alois Fürstner

Max-Planck-Institut für Kohlenforschung, Kaiser-Wilhelm-Platz 1, D–45470 Mülheim an der Ruhr, Germany

Selected total syntheses of structurally diverse natural products are summarized which are invariably based upon the elaboration of sugar-derived building blocks by olefin metathesis. The chosen targets are conduritol F and congeners, the marine lipidic lactone ascidiatrienolide, the phytotoxic agents herbarumin and pinolidoxin, as well as the complex resin glycosides tricolorin and woodrosin. These examples allow to deduce some of the strategic advantages of metathesis in advanced organic synthesis.

Introduction

The advent of well defined metal alkylidene complexes such as **1**-**6**, that are able to catalyze olefin metathesis reactions with high efficiency yet tolerate a wide range of polar functional groups, has had a profound impact on all branches of organic synthesis (*1*), with carbohydrate chemistry being no exception to the rule (*2*). Although the following account does not intend to provide a comprehensive coverage of this prosperous field of research, many advantages associated with metathetic conversions in general and with ring closing alkene metathesis (RCM) in particular will become evident from the case studies outlined below. They provide a brief survey of our recent efforts on the elaboration of sugar-derived building blocks into natural products of different complexity by exploiting some of the unique chemical features of RCM (*3*).

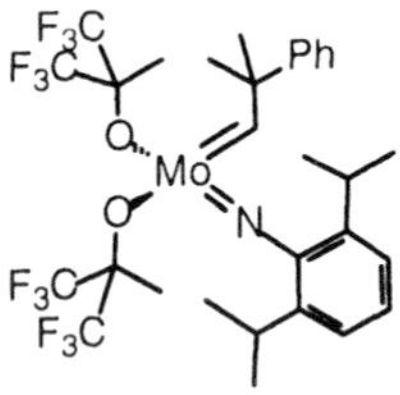

1 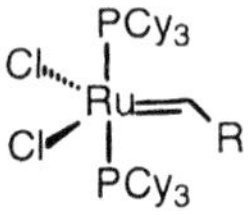 **2** (R = Ph, CH=CPh_2)

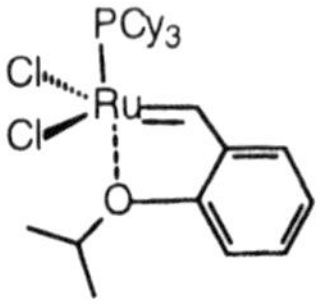

3 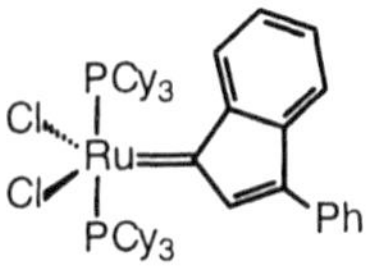 **4** 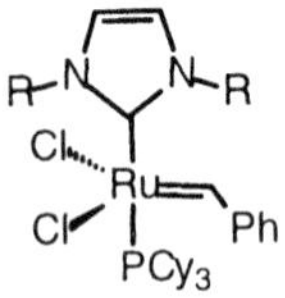 **5** 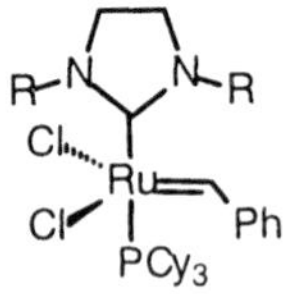 **6**

Conduritols: Vignette on Catalyst Activity

Polyhydroxylated cyclohexene derivatives such as conduritol A-F (**7**-**12**) are important secondary messengers eliciting a wide range of biological responses and have also been widely used as starting materials for advanced organic synthesis (*4*). RCM opens a particularly concise entry into this class of compounds, not least because of the ready availability of the required cyclization precursors from abundant monosaccharides (*5,6*).

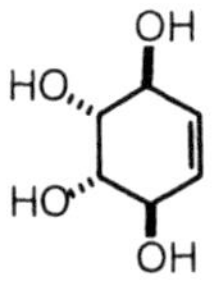

Conduritol A (**7**) 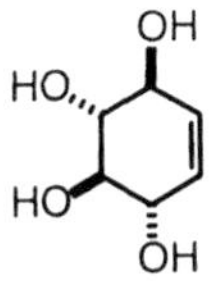 Conduritol B (**8**) 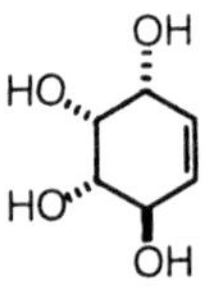 Conduritol C (**9**)

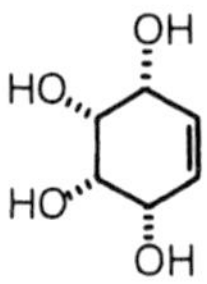

Conduritol D (**10**) 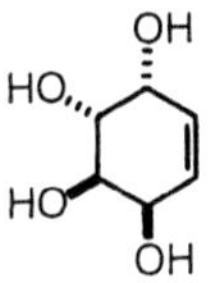 Conduritol E (**11**) 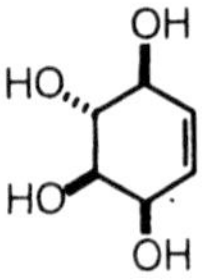 Conduritol F (**12**)

This is exemplified by the preparation of diene **18** from D-glucitol **13** depicted in Scheme 1 (*5*). Standard protecting group manipulations afford diol **16** in excellent overall yield on a large scale, which was subjected to a Swern oxidation to afford dialdehyde **17**. Due to the lability of this compound, however, the subsequent conversion into diene **18** turned out to be more delicate than anticipated and was best carried out using *Tebbe*'s reagent [$Cp_2Ti(\mu\text{-}Cl)(\mu\text{-}CH_2)AlMe_2$] in THF/pyridine as the solvent.

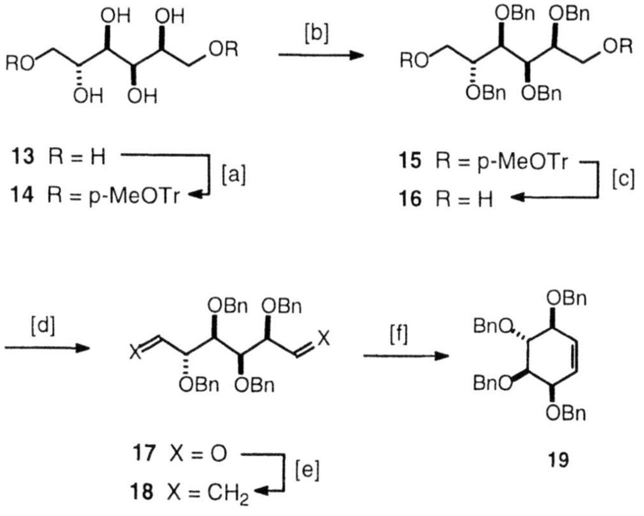

OH OH
RO OR
OH OH
[b]
OBn OBn
RO OR
OBn OBn

13 R = H
14 R = p-MeOTr
[a]

15 R = p-MeOTr
16 R = H
[c]

[d]
OBn OBn
X X
OBn OBn
[f]
OBn
BnO
BnO
OBn

17 X = O
18 $X = CH_2$
[e]

19

*Scheme 1. [a] p-Methoxytrityl chloride, DMAP, pyridine, r.t., 85%; [b] BnBr, NaH, THF, 96%; [c] H_2SO_4 cat., MeOH/CH_2Cl_2, 0°C, 87%; [d] DMSO, oxalyl chloride, NEt_3, CH_2Cl_2, –78°C→r.t., quant.; [e] $Cp_2Ti(\mu\text{-}Cl)(\mu\text{-}CH_2)AlMe_2$, THF/pyridine, –40°C→r.t., 42%; [f] catalyst **2** (5 mol%), 60h, 32% (GC);* or: *catalyst **1** (5 mol%), 1h, 92%;* or: *catalyst **5** (R = mesityl), 2h, 89%.*

The cyclization of this key intermediate to the fully protected conduritol F derivative **19** showcases the different performance of the standard metathesis catalysts. Despite the excellent track record of the original *Grubbs* benzylidene carbene complex **2** (*7*) for the cyclization of 6-membered rings, compound **18** reacts poorly with this particular catalyst in refluxing CH_2Cl_2, leading to only 32% conversion after 60h reaction time. This reluctance is likely caused by the preference of diene **18** to adopt a zig-zag-conformation holding the olefin units far apart.

Importantly, however, the molybdenum alkylidene complex **1** developed by *Schrock* (*8*) exhibits much higher reactivity and converts this substrate into the corresponding cycloalkene **19** in less than 1h in 92% isolated yield. Similar results were obtained with the 'second-generation' ruthenium carbene **5** (R = mesityl) bearing one imidazol-2-ylidene ligand (*9*), which delivered the desired product in similar yield after 2h reaction time. In view of the significantly higher stability of **5** toward air and moisture, this reagent has clear-cut advantages in preparative terms over the quite sensitive molybdenum species **1** which must be handled using rigorous Schlenck techniques. While it is now widely accepted that ruthenium complexes such as **5** or **6** with N-heterocyclic carbenes in their ligand sphere rival *Schrock*'s catalyst in many respects (*1*), this approach to conduritol F (and several other members of this series) was one of the first reports providing a direct and quantitative comparison of the available catalyst systems (*5*).

Herbarumin and Pinolidoxin: Stereochemical Control by Catalyst Tuning

Although RCM opens access to carbo- and heterocycles of any size ≥ 5, the synthesis of 8-11-membered rings still poses considerable challenges due to the fact that ring strain predisposes such systems for the reverse process, that is, for ring-opening metathesis (ROM) or ring opening metathesis polymerization (ROMP) (*1*). Therefore the total synthesis of herbrumin I **20** and II **21** as well as pinolidoxin **22**, three potent herbicidal agents isolated from the fungi *Phoma herbarum* (*10*) and *Ascochyta pinoides* (*11*), respectively, were far from routine. Since neither the relative nor the absolute stereochemistry of **22** had been established at the outset of our project, it was necessary to gain access to all possible isomers for comparison of their data with those of the natural product (*12,13*).

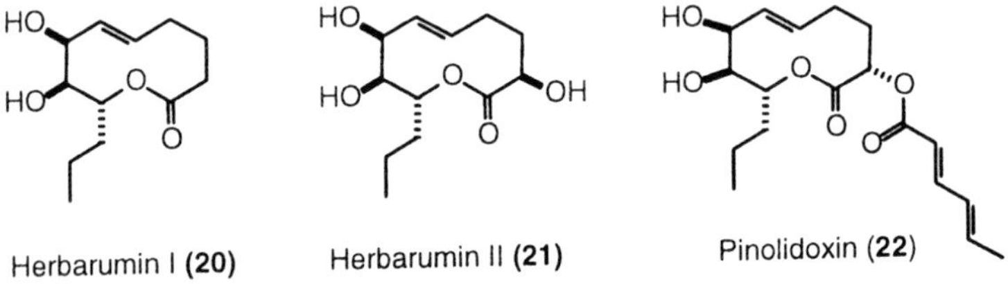

HO HO O O — Herbarumin I **(20)**

HO HO O O OH — Herbarumin II **(21)**

HO HO O O O O — Pinolidoxin **(22)**

In addition to the concerns about the stability of 10-membered cycloalkenes under the reaction conditions, one has to keep in mind that RCM tends to give mixtures of the (*E*)- and (*Z*)-configured products when applied to compounds of this size (*1*). Because a reliable catalyst allowing to control the configuration of the newly formed double bond has yet to be found (*14*), recourse was taken to a synthesis plan which allows to confer bias to cyclization by external control elements.

Specifically, we reasoned that a cyclic protecting group for the *vic*-diol unit of the targets should align the olefinic subunits of the required diene precursors in a cyclization-friendly conformation. Moreover, it is reasonable to expect that the (*E*)- and the (*Z*)-isomers of the resulting products show significantly different ring strain because the olefins are embedded into a rigid bicyclic structure; this enthalpic distinction might translate into good stereoselectivity during RCM. In fact, semiempirical calculations for the isopropylidene acetal of herbarumin I indicated that the (*Z*)-olefin is ca. 3.5 kcal mol^{-1} more stable than its (*E*)-configured counterpart (*13*). Hence, RCM under kinetic control is called for en route to products **20-22** which invariably incorporate an (*E*)-alkene unit. Therefore ruthenium complexes bearing two phosphine ligands such as **2** or **4** should be the catalysts of choice, whereas the more reactive 'second generation' ruthenium complexes **5** or **6** bearing an N-heterocyclic carbene ligands must be avoided as they tend to equilibrate the products initially formed and hence favor the thermodynamically more stable isomer (*16,17*).

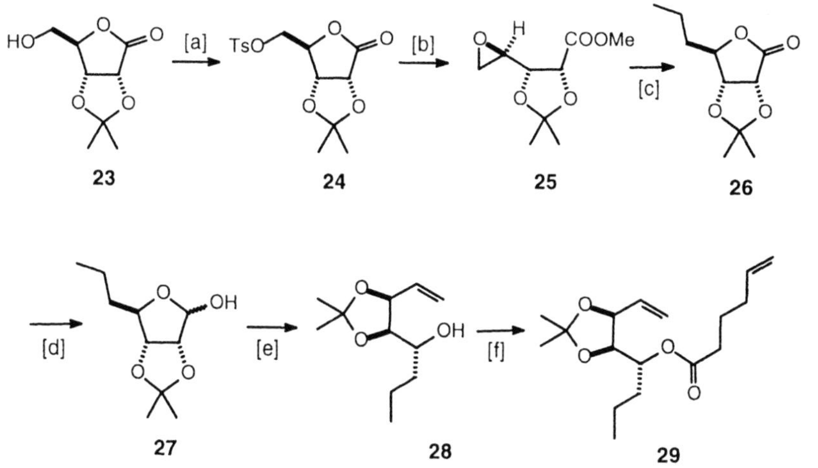

Scheme 2. [a] Tosyl chloride, pyridine, –25°, 77%; [b] NaOMe, THF, 0°C → r.t., 62%; [c] EtMgBr, CuBr·Me_2S, THF, –78°C → r.t., 60%; [d] Dibal-H, CH_2Cl_2, –78°C, 97%; [e] $Ph_3P{=}CH_2$, quinuclidine cat., THF, reflux, 62-77%; [f] 5-hexenoic acid, DCC, DMAP, CH_2Cl_2, 84%.

This analysis turned out to be in excellent agreement with the experimental results (*13*). The required chiral building block for the synthesis of the herbarumins was prepared from **D**-ribonolactone as shown in Scheme 2. Thus, acetonide **23** was converted on a multigram scale into the corresponding tosylate **24**. Exposure to MeONa in THF results in a transesterification with release of an alkoxide at O-4 which displaces the adjacent tosyl group to form the terminal epoxide **25**. This compound was then reacted with EtMgBr in the presence of catalytic amounts of $CuBr{\cdot}Me_2S$ to provide lactone **26**, which was reduced with Dibal-H to afford the corresponding lactol **27**. Subsequent Wittig reaction delivered the desired triol segment **28** in suitably protected form which was esterified with 5-hexenoic acid to give the cyclization precursor **29** in excellent overall yield (Scheme 2) (*12,13*).

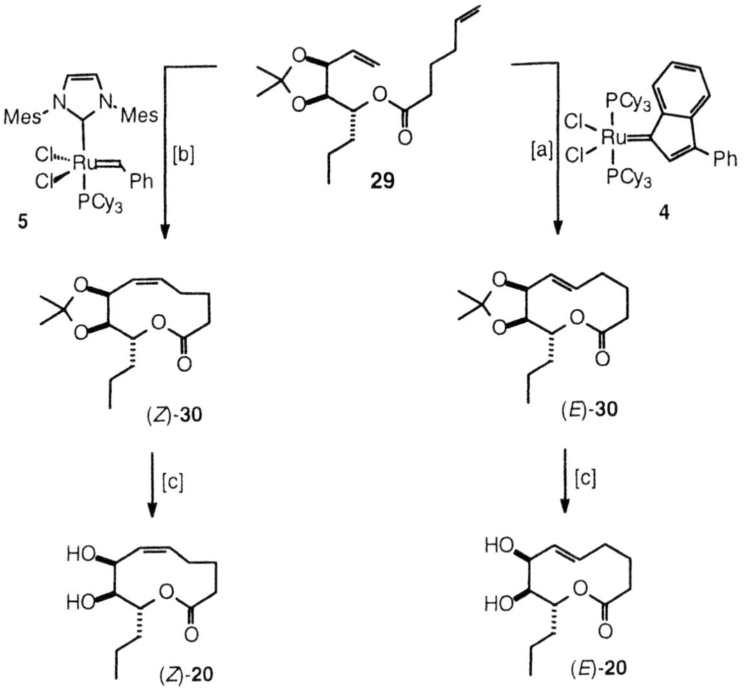

*Scheme 3. [a] Complex **4** cat., CH_2Cl_2, reflux, 69%; [b] complex **5** cat., CH_2Cl_2, reflux, 86%; [c] aq. HCl, THF, 50°C, 47% ((Z)-**20**), 90% ((E)-**20**).*

The outcome of the RCM experiments employing different catalysts was consistent with the analysis outlined above (Scheme 3). Exposure of diene **29** to the ruthenium indenylidene complex **4** (*15*), a readily available alternative to the classical *Grubbs* catalyst, afforded the desired (*E*)-configured lactone (*E*)-**30** in 69% isolated yield, together with only 9% of the corresponding (Z)-isomer. Importantly, the *E:Z* ratio does not evolve with time, indicating that the observed stereoselectivity is likely kinetic in origin. In contrast, the use of the 'second generation' metathesis catalyst **5** (*9*) led to the opposite stereochemical outcome, furnishing (*Z*)-**30** in 86% yield. Because of its higher overall activity, complex **5** and congeners are able to isomerize cycloalkenes during the course of the reaction and hence enrich the product in the thermodynamically more stable isomer (*16,17*). This particular application may therefore be considered the first example of a *rationally designed synthesis of both possible stereoisomers of a given cycloalkene* by adapting the activity of the RCM catalyst to the structural peculiarities of the chosen target (*12,13*). Finally, cleavage of the acetal group in **30** furnished herbarumin I (*E*)-**20** in high yield.

Along similar lines, we secured access to herbarumin II **21** and have also been able to prepare a comprehensive set of stereoisomers of pinolidoxin (*13*). Among them, compound **22** turned out to be identical to the natural product, thereby correcting the relative and absolute stereostructure of this herbicidal agent which had been misassigned in the original publications (*11*).

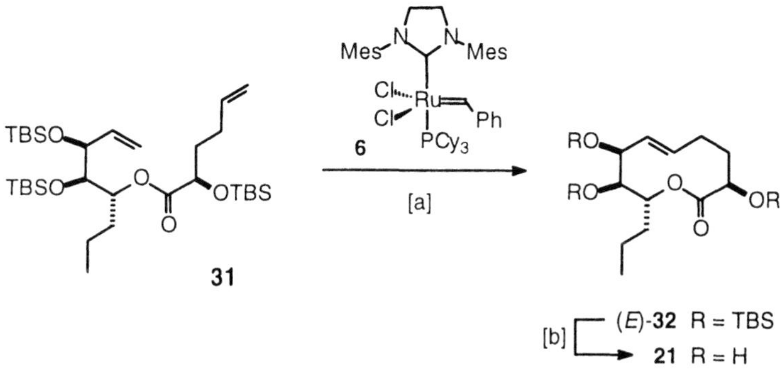

Scheme 4. [a] catalyst ***6*** *(20 mol%),* CH_2Cl_2*, reflux, 8h, 85%; [b]* Bu_4NF*, THF, 1.5h, quant.*

Very recently, an alternative approach to herbarumin II **21** via RCM has been published which nicely corroborates the concept of stereocontrol in constrained systems by catalyst tuning (Scheme 4) (*18*). Instead of using an isopropylidene protecting group for the *vic*-diol unit of the cyclization precursor, *Ley* et al. chose substrate **31** bearing bulky *tert*-butyl-dimethylsilyl (TBS) ether moieties. *Macromodel* studies had suggested that this pattern renders the (*E*)-configured cycloalkene more stable by 2.5 kcal mol^{-1}. This difference to the isopropylidene series is explained by the conformation of the 10-membered ring of **32** in which the bulky substituents adopt pseudo-diaxial orientations minimizing steric clash. As a consequence – and in striking contrast to the isopropylidene series – 'second generation' ruthenium carbene complexes should now be the catalysts of choice en route to **21**. In fact, exposure of diene **31** to **6** (20 mol%) in refluxing CH_2Cl_2 delivered (*E*)-**32** in 85% yield, whereas the ruthenium indenylidene species **4** produced the now kinetically favored (*Z*)-isomer as the major compound (*E*:*Z* = 1:2) (*18,19*).

Ascidiatrienolide A and the Didemnilactones

Crude extracts of the ascidian *Didemnum candidum* exhibit strong inhibitory effects against phospholipase A_2 in vitro. A search for the active component led to the discovery of ascidiatrienolide A **33** (*20*), an eicosatetraene derivative that is closely related to the didemnilactones **34-36** derived from the tunicate *Didemnum moseley* (*21*). The latter are endowed with high affinity to the leukotriene B_4 receptor of human polymorphonuclear leucocyte membrane fractions. These fatty acid derivatives of marine origin provided yet another opportunity to validate our strategy for (*E,Z*)-control outlined in the previous chapter based upon proper matching of the reactivity of the metathesis catalyst with the conformational preferences of a constrained substrate (*22*).

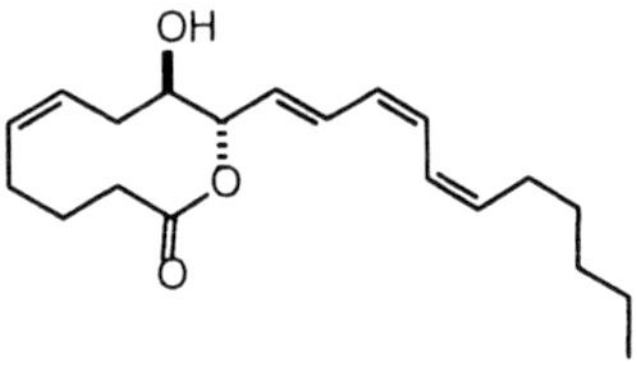

Ascidiatrienolide A (**33**)

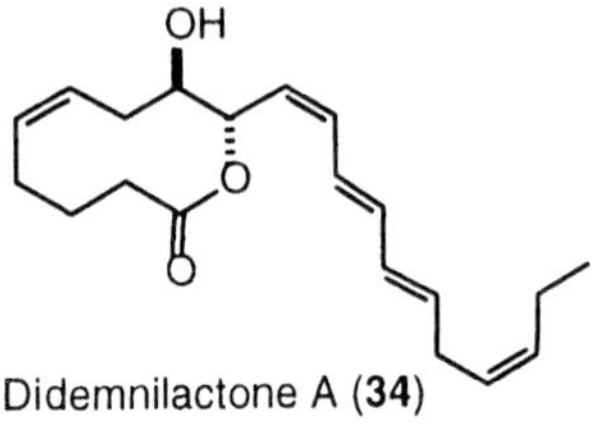

Didemnilactone A (**34**)

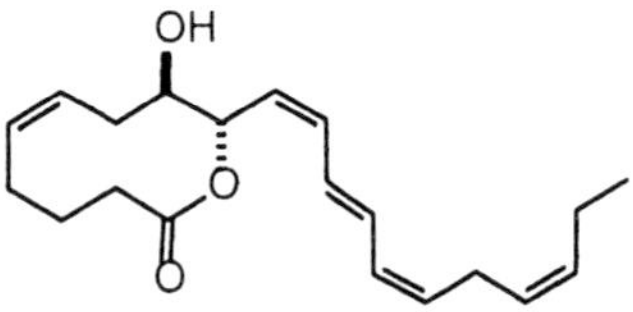

Didemnilactone B (**35**)

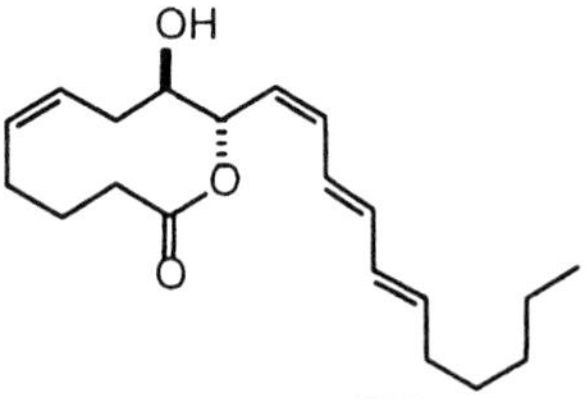

Neodidemnilactone (**36**)

The common nonenolide core of ascidiatrienolide and the didemnilactones has been formed as shown in Scheme 5 (*22*). Specifically, a tin-mediated, ultrasound-promoted addition of allyl bromide to unprotected glyceraldehyde **37** furnished a mixture of the corresponding homoallyl alcohols **38** which were subjected to acetalization and esterification with 5-hexenoic acid under standard conditions. At that stage, the major *syn*-isomer **40** can be purified by flash chromatography. Exposure of this diene to a refluxing solution of the ruthenium indenylidene complex **4** provided the cycloalkene **41** in excellent overall yield with a *E:Z* ratio of 3.5:1 (*22*). In contrast, the use of the 'second generation' catalyst **5** resulted in a reversed stereochemical outcome, affording the desired (*Z*)-isomer as the major product. This stereochemical divergence corroborates the concept of catalyst tuning discussed above.

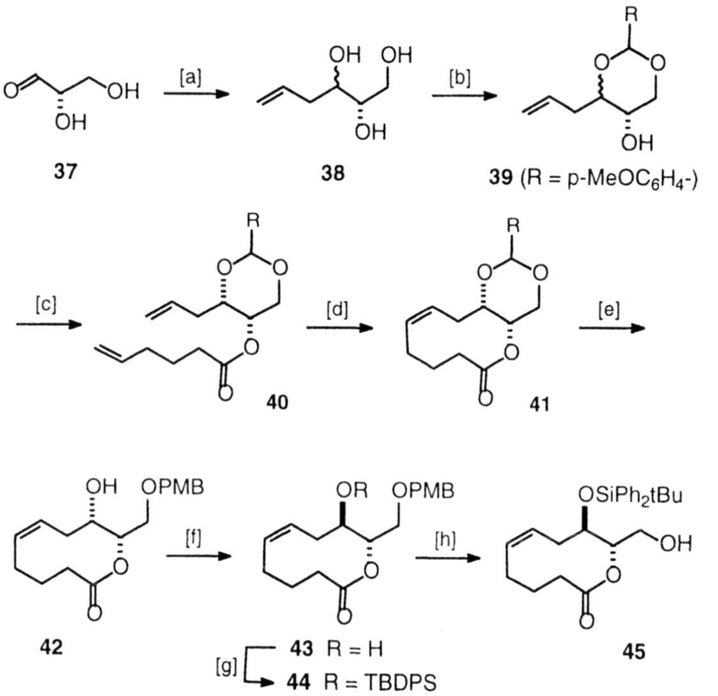

Scheme 5. [a] Allyl bromide, Sn, ultrasound, 98%; [b] p-$MeOC_6H_4CH(OMe)_2$*, camphorsulfonic acid cat.,* CH_2Cl_2*, 67%; [c] 5-hexenoic acid, DIC, DMAP,* CH_2Cl_2*, 40% (pure stereoisomer); [d] complex* **5** *(R = mesityl),* CH_2Cl_2*, reflux, 79% E:Z = 1:2.8; [e]* $NaBH_3CN$*, HCl, THF, 71%; [f] (i) chloroacetic acid,* PPh_3*, DEAD, THF, 98%; (ii)* K_2CO_3*, MeOH, 59%;[g] TBDMSOTf, 2,6-lutidine,* CH_2Cl_2*, 81%; [h] DDQ,* CH_2Cl_2/H_2O*, 98%.*

Compound **41** thus formed was elaborated into the common nonenolide core of ascidiatrienolide A and the didemnilactones by reductive opening of the *p*-methoxybenzylidene acetal with $NaBH_3CN$ in the presence of ethereal HCl, followed by a virtually quantitative Mitsunobu reaction of the resulting alcohol

42 with chloroacetate. Standard protecting group manipulations as shown in Scheme 5 furnished product **45** which constitutes the key intermediate of a previous total synthesis of ascidiatrienolide **33** (*20b*). By adapting the established chain elongation protocol, compound **45** could also be elaborated into the individual members of the didemnilactone series (*22*).

Tricolorin A and G

Plants belonging to the morning glory family (*Convolvulaceae*) are rich sources of alkaloids and resin glycosides and have been extensively used in traditional medicine as herbal remedies for various diseases. Resin glycosides are amongst the most complex glycolipids known to date (*23*). They combine a complex oligosaccharide backbone with (11*S*)-hydroxyhexadecanoic acid ("jalapinolic acid") as characteristic aglycon that is preserved in virtually all members of this series. The latter is frequently tied back to form a macrolactone ring spanning two or more sugar units of the backbone. Further carboxylic acids may complement the peripheral acylation pattern.

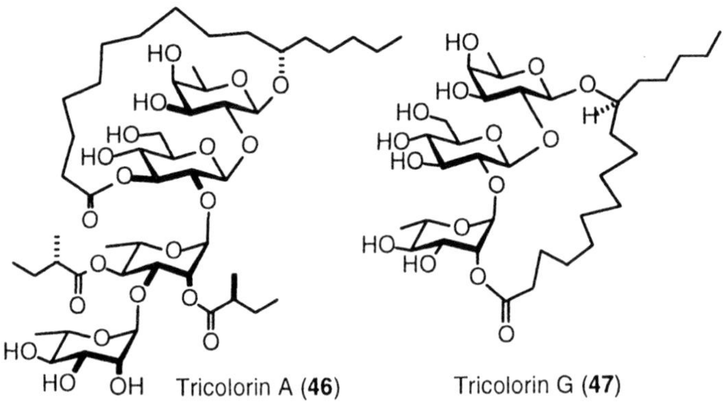

Tricolorin A (**46**)   Tricolorin G (**47**)

Tricolorin A (**46**) and G (**47**) are prototype members of this class of amphiphilic glycoconjugates (*24*). They constitute the allelochemical principles of *Ipomoea tricolor* Cav., a plant used in traditional agriculture in Mexico as a cover crop to protect sugar cane against invasive weeds. Their molecular mechanism of action likely involves the inhibition of the H^+-ATPase of the plasma membrane, an enzyme that plays a crucial role in plant cell physiology. Moreover, **46** acts as a natural uncoupler of photophosphorylation in spinach chloroplasts. This compound also displays general cytotoxicity against several cancer cell lines and is able to antagonize phorbol ester binding to protein kinase C (*23,24*).

Although efficient total syntheses of **46** via macrolactonization had been published (*25*), it was expected that an alternative approach based on RCM for the formation of the large ring might be beneficial in certain respects. Specifically, RCM has the inherent advantage that systematic variations of the ring size and hence of the lipophilicity of the compound are easy to accomplish (*26,27*).

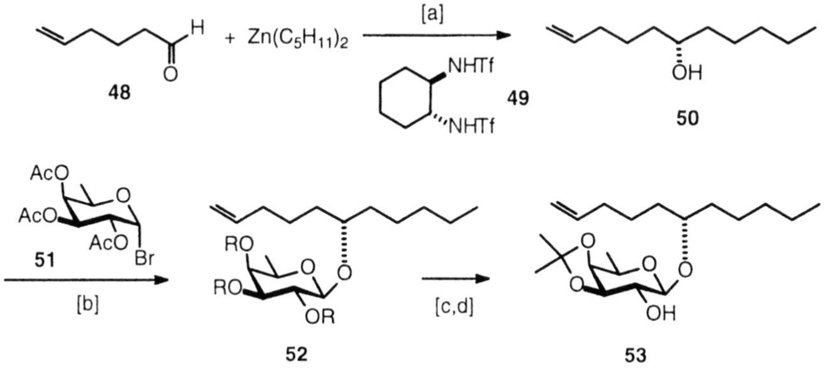

[a]
+ $Zn(C_5H_{11})_2$
NHTf
NHTf
48
49
50
OH
AcO
AcO
AcO
51
Br
[b]
RO
RO
OR
52
[c,d]
OH
53

*Scheme 6. [a] $Ti(OiPr)_4$, ligand **49** (cat.), toluene, 86% (ee > 99%); [b] $AgNO_3$ on silica/alumina, MS 3Å, CH_2Cl_2, −10°C, 69%; [c] KOMe cat., MeOH,; [d] 2,2-dimethoxypropane, p-TsOH·H_2O cat, acetone, 91% (over both steps).*

The synthesis of the required fucose building block is shown in Scheme 6. Enantioselective addition of dipentylzinc to 5-hexenal **48** in the presence of $Ti(O\text{-}iPr)_4$ and catalytic amounts of ligand **49** proceeded smoothly on a multigram scale, providing the (*S*)-configurated alcohol **50** in good yield and excellent enantiomeric purity (ee ≥ 99%). Its glycosylation with the known tri-O-acetyl-α-**D**-fucopyranosyl bromide **51** in the presence of $AgNO_3$ on silica/alumina gave compound **52** in 69% yield, which was deacetylated and subsequently converted into the isopropylidene acetal **53** under standard conditions (*26*).

Reaction of this fucose derivative with the glucosyl donor **54** (α:β ≈ 2:1) in the presence of catalytic amounts of $BF_3{\cdot}Et_2O$ cleanly provided the desired β-configurated disaccharide **55**. Deprotection followed by acylation of the resulting diol **55a** with 6-heptenoic acid in the presence of DCC and DMAP proceeded selectively at the O-3 position, thus affording ester **56** as the only product in 71% yield (Scheme 7) (*26,27*).

In line with our previous experiences, diene **56** readily cyclized to the desired 19-membered ring **57** on reaction with the ruthenium carbene **2** (5 mol%) in refluxing CH_2Cl_2. The fact that neither the free hydroxyl group nor any other functionality in the substrate interfere with RCM illustrates the excellent compatibility and selectivity of the *Grubbs* catalyst (*7*). Hydrogenation of the crude cycloalkene (*E/Z*-mixture) thus obtained afforded the desired disaccharide **57** in 77% yield. The elaboration of this compound into tricolorin A **46** can be achieved according to literature procedures (*25*).

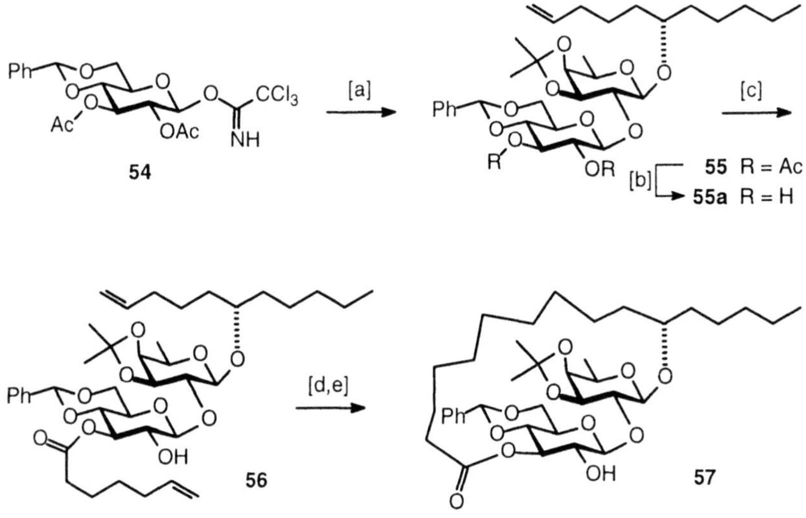

Scheme 7. [a] Compound **53**, *$BF_3{\cdot}Et_2O$ cat, CH_2Cl_2/n-hexane, −20°C, 82%; [b] KOMe cat., MeOH, 71%; [c] 6-heptenoic acid, DCC, DMAP, CH_2Cl_2; [d] catalyst* **2** *(5 mol%), CH_2Cl_2, reflux; [e] H_2 (1 atm), Pd/C (5 mol%), EtOH, 77% (over both steps).*

The inherent flexibility of the RCM approach was illustrated by the synthesis of several analogues differing from tricolorin A in their ring size and hence lipophilicity (Scheme 8). For this purpose, diol **55a** was reacted with unsaturated acid derivatives other than 6-heptenoic acid and the dienes **58** and **59** thus formed were processed as described above. This allowed us to access compound **60** missing two of the CH_2-groups in the macrocyclic loop of tricolorin A as well as compound **61** featuring a considerably expanded lactone moiety (*26*).

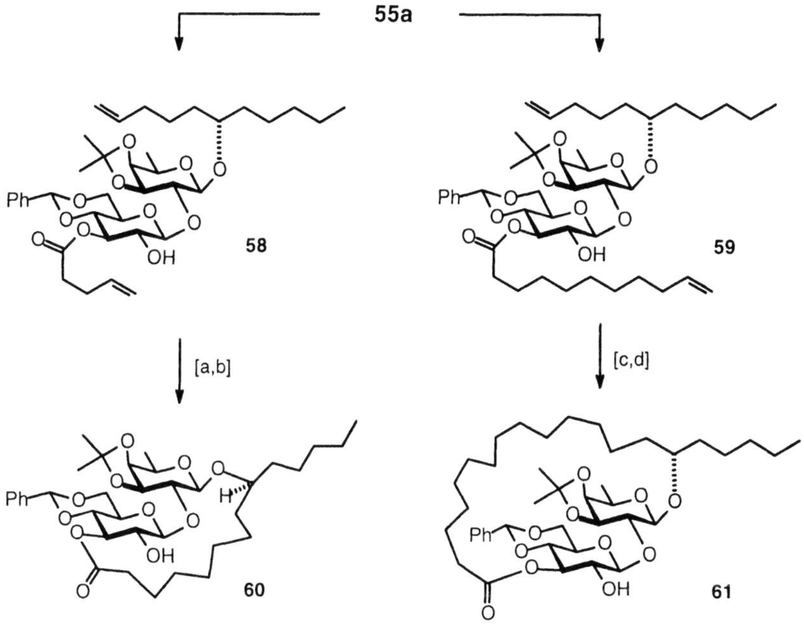

*Scheme 8. [a] Catalyst **2** (20 mol%), CH_2Cl_2, reflux; [b] H_2 (1 atm), Pd/C (5 mol%), EtOH, 76% (over both steps); [c] catalyst **2** (10 mol%), CH_2Cl_2, reflux; [b] H_2 (1 atm), Pd/C (5 mol%), EtOH, 76% (over both steps).*

Along similar lines, diol **55a** was also used as the key intermediate for the first total synthesis of tricolorin G **47** in which jalapinolic acid spans a trisaccharidic backbone (Scheme 9). For this purpose, a properly functionalized rhamnose unit was attached to the *less* reactive C2'' hydroxyl group of diol **55a** prior to the crucial RCM reaction. This was ensured by blocking the more reactive site as a benzyl ether followed by reaction of the resulting product with trichloroacetimidate **62** in the presence of $BF_3 \cdot Et_2O$ as the catalyst. Exchange of the acetyl group on the rhamnose for the corresponding 6-heptenoic acid ester delivered diene **63** as the substrate for the subsequent RCM reaction which was again highly productive. Saturation of the resulting cycloalkene to give **64** was followed by global deprotection providing tricolorin G **47** in excellent overall yield (*26*). Once again, the flexibility of the chosen approach was highlighted by the preparation of several analogues of this intricate glycolipid without undue preparative efforts (*26*).

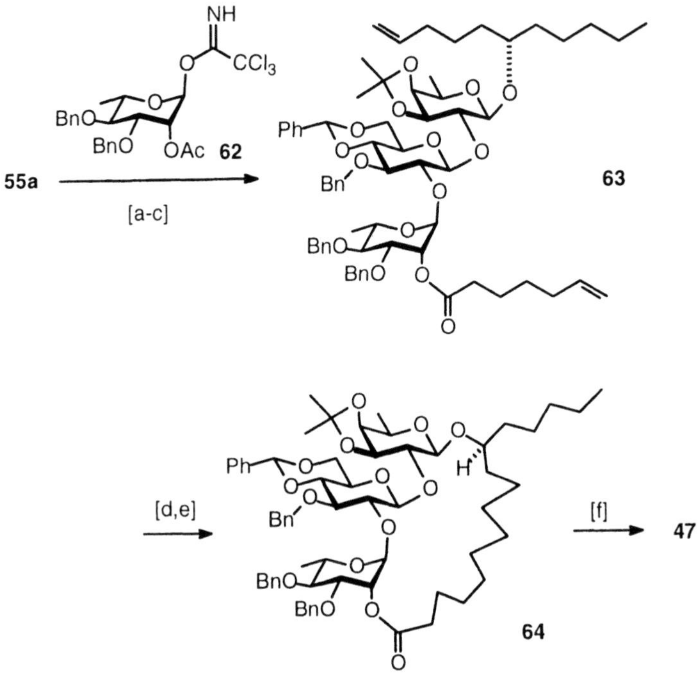

*Scheme 9. [a] $BF_3{\cdot}Et_2O$ cat., CH_2Cl_2/n-hexane, –20°C, 53%; [b] KOMe, MeOH; [c] 6-heptenoic acid, DCC, DMAP cat., CH_2Cl_2, 84% (over both steps); [d] complex **2** (10 mol%), CH_2Cl_2, reflux; [e] H_2 (1 atm), $RhCl(PPh_3)_3$ cat., EtOH, 93% (over both steps); [f] (i) F_3CCOOH, CH_2Cl_2; (ii) H_2 (1 atm), Pd/C, MeOH, F_3CCOOH cat., 49%.*

Woodrosin

Woodrosin I (**65**) is an ether-insoluble glycolipid isolated from the stems of *Ipomoea tuberosa* L. (*Merremia tuberosa* (L.) Rendle) commonly called "woodrose" after the shape of its dried calyx (*28*). Its lactone moiety spans four glucose units to form a 27-membered macrocycle. Further challenges arise from the additional acylation pattern at the periphery of the branched and hence highly hindered pentasaccharide perimeter of this glycoconjugate.

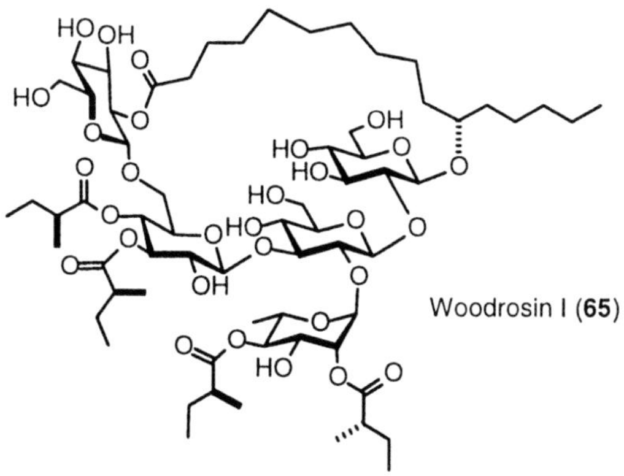

Woodrosin I (**65**)

Despite of these demanding structural features, we were able to complete a total synthesis of this intricate target using RCM as the means to forge the macrocyclic ring (*29,30*). The cyclization precursor was prepared in a highly convergent manner from the building blocks shown in Scheme 10 (30*a*).

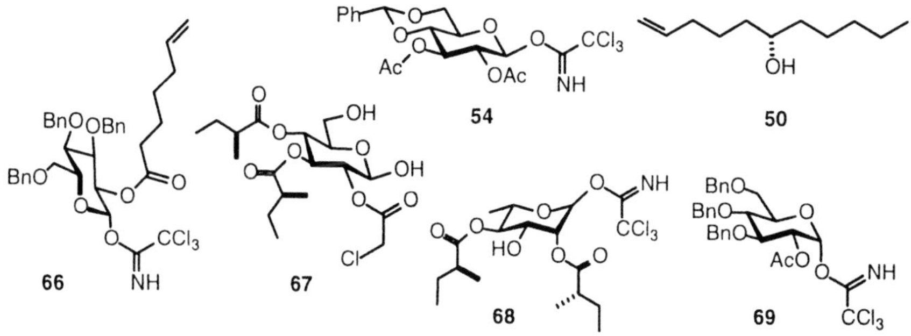

Scheme 10. Building blocks required for the total synthesis of woodrosin I (30a).

The assembly of these fragments to the oligosaccharide perimeter of woodrosin I, however, turned out to be far more difficult than anticipated and required a careful optimization of events (Scheme 11) (*29,30*). Among the different possibilities that might be envisaged, the ultimately successful route started off with a reaction of (6*S*)-undecenol **50** with trichloroacetimidate **69** to give glucoside **70**, which was deacetylated and then used as glycosyl acceptor in a reaction with donor **54** to give product **71** in high overall yield.

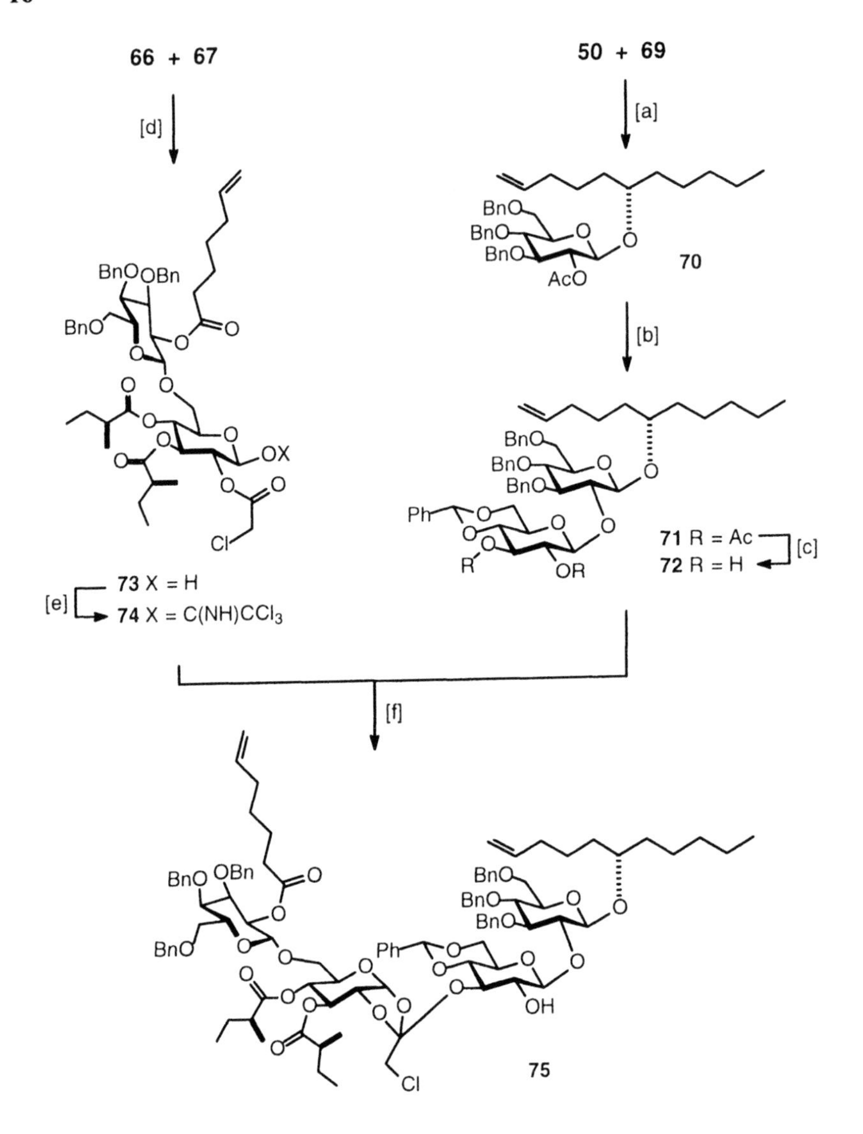

*Scheme 11. Assembly of the metathesis precursor en route to woodrosin I. [a] $BF_3{\cdot}Et_2O$, CH_2Cl_2/hexane, –20°C, 68%; [b] (i) NaOMe cat., MeOH, quant.; (ii) compound **54**, $BF_3{\cdot}Et_2O$, CH_2Cl_2/hexane, –20°C, 82%; [c] NaOMe cat., MeOH, 88%; [d] $BF_3{\cdot}Et_2O$, CH_2Cl_2, –20°C, 86%; [e] Cl_3CCN, Cs_2CO_3, CH_2Cl_2, 54%; [f] TMSOTf cat., CH_2Cl_2, –20°C, 84%.*

The other disaccharide building block was formed from trichloroacetimidate **66** and the 1,6-unprotected compound **67** by a fully regioselective glycosidation event. Since the anomeric center of the resulting product **73** is unmasked, this compound was directly amenable to the preparation of the next donor **74**. The smooth reiterability of the trichloroacetimidate method by using 1,ω-di-unprotected synthons represents a promising but as yet largely unexplored aspect of this methodology.

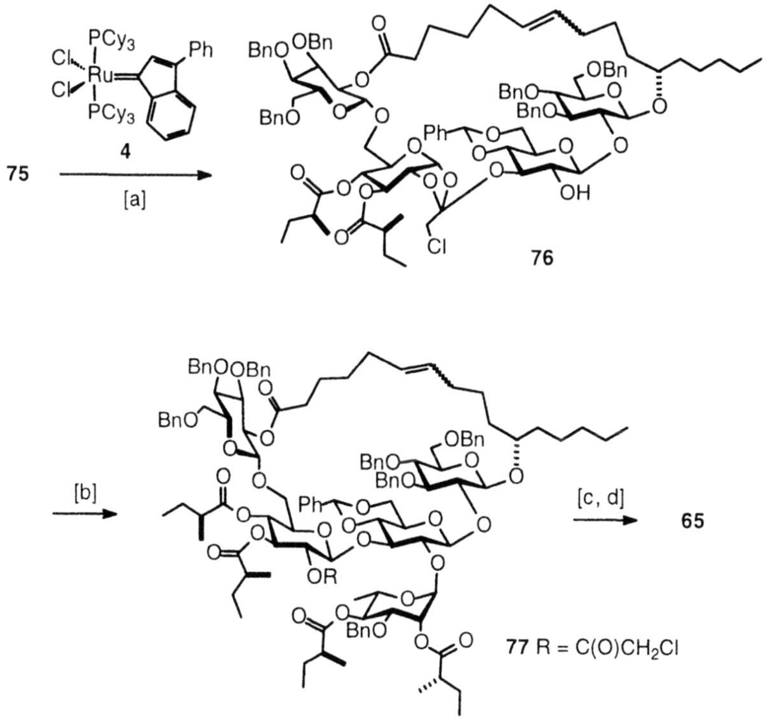

*Scheme 12. Completion of the total synthesis of woodrosin I. [a] complex **4** cat., CH_2Cl_2, reflux, 94%; [b] donor **68**, TMSOTf cat., Et_2O, 0°C, 60%; [c] hydrazinium acetate, DMF, –10 → 0°C; [d] H_2, Pd/C, MeOH, 84% (over both steps).*

The subsequent reaction of compound **72** with trichloroacetimidate **74** turned out to be highly regioselective but delivered orthoester **75** rather than the desired β-glycoside as the only reaction product despite considerable experimentation with different promoters and solvents (Scheme 11). Moreover, attempted rhamnosylation of **75** on treatment with donor **68** failed due to the severe shielding of the remaining hydroxyl group in the bay area of the orthoester. Therefore, the completion of the sugar backbone had to be postponed until after the RCM reaction in the believe that the trajectory to this hidden OH-group might be less narrow after forging the macrocyclic motif as suggested by molecular modeling.

Diene **75** cyclized with exceptional ease on exposure to the ruthenium indenylidene complex **4** previously developed in our laboratories (*15*), providing cycloalkene **76** in virtually quantitative yield. It was gratifying to note that slow addition of donor **68** to a solution of **76** and TMSOTf in Et_2O at 0°C did not only allow to introduce the yet missing rhamnose unit to the innately unreactive 2''-OH site but also induced the concomitant rearrangement of the adjacent orthoester to the required β-glycoside. This favorable coincidence illustrates the maturity of the trichloroacetimidate method (*31*) and represents one of the most advances examples of Kochetkov's orthoester protocol (*32*) reported to date. Global deprotoection of product **77** thus formed completed the first total synthesis of woodrosin I **65** (Scheme 12) (*29,30*).

Conclusions

Although the case studies summarized above can only cover a small segment of metathesis chemistry, they illustrate some of the strategic advantages of this transformation. Olefin metathesis has evolved within a few years into a reliable tool that can be safely implemented even into complex settings (*1*). Chemists have gained enough confidence in this reaction to subject elaborate and highly valuable materials to it. The most notable features of metathesis as it stands today may be summarized as follows:

1. As illustrated by the syntheses of woodrosin or the tricolorins, metathesis allows for the disconnection of complex targets at *remote* sites, e.g. within an unfunctionalized hydrocarbon chain. This aspect is best appreciated when seen in the context of established retrosynthetic analyses which are uniformly guided by the polarity induced by the substituents of a given target and therefore invariably opt for disconnections at or close to the functional groups. Therefore one must conclude that RCM *is complementary to the existing arsenal and expands beyond conventional retrosynthetic logic.*

2. RCM opens access of carbo- and heterocycles of any ring size ≥ 5, including even the kinetically and thermodynamically most

handicapped medium-large systems, and is particularly efficient when applied to the macrocyclic series.

3. Our approaches to herbarumin, pinolidoxin, and ascidiatrienolide illustrate for the first time that proper planning of such syntheses allows for a rational approach to cycloalkenes with defined configurations at the newly formed double bond (*13,22*).

4. The superb chemoselectivity of the available catalysts alleviates many issues associated with protecting group chemistry plaguing more conventional methods for C-C-bond formation; even substrates with free –OH functions or other seemingly reactive sites can be processed without difficulty.

5. As a result of the minimal substrate protection required, metathesis allows to design synthesis routes that are (unprecedentedly) short and therefore meet the demands for an "economy of steps" as a prime criterion of contemporary organic synthesis (*33*).

6. Metathesis is inherently flexible. As a result, this transformation allows, inter alia, for a rapid chemistry-driven evaluation of structure/activity profiles and qualifies for applications in combinatorial and diversity oriented synthesis; this is particularly true when combined with suitable post-metathesis transformations exploiting diverse alkene reactivity.

Although several aspects of alkene metathesis need further improvement or remain to be solved, there is no doubt that this transformation has a profound impact on all sub-disciplines of modern organic chemistry and will continue to shape preparative carbohydrate chemistry and natural product synthesis in the future.

Experimental

Compound (Z)-30. A solution of diene **29** (68.4 mg, 0.231 mmol) and the ruthenium complex **5** (19.5 mg, 0.023 mmol) in CH_2Cl_2 (100 mL) is refluxed for 8 h until TLC shows complete conversion of the substrate. After quenching the reaction with ethyl vinyl ether (0.5 mL), all volatiles are removed in vacuo and the residue is purified by flash chromatography (pentane/Et_2O, 10:1) to afford compound (Z)-**30** as a colorless syrup (53 mg, 86%). $[\alpha]_D^{20} = -83.3$ (c 0.6, CH_2Cl_2). IR: 3018, 1743, 1659, 1210 cm^{-1}. ^1H NMR (600 MHz, 303 K, $CDCl_3$): δ 5.40 (ddd, *J* = 11.6, 4.2, 1.0 Hz, 1H), 5.34 (ddd, *J* = 11.4, 9.5, 2.0 Hz, 1H), 4.92 (dd, *J* = 9.5, 6.2 Hz, 1H), 4.64 (ddd, *J* = 9.9, 7.7, 3.6 Hz, 1H), 4.18 (dd, *J* = 9.9, 6.1 Hz, 1H), 2.28 (dddd, *J* = 11.7, 4.5, 3.5, 0.6 Hz, 1H), 2.23 (ddd, *J* = 13.7, 11.5, 1.4 Hz, 1H), 2.15 (m, 1H), 2.13 (m, 1H), 1.91 (m, 1H), 1.76 (m, 1H), 1.75

(m, 1H), 1.56 (m, 1H), 1.45 (s, 3H), 1.37 (m, 2H), 1.33 (s, 3H), 0.89 (t, J = 7.3 Hz, 3H). ^{13}C NMR (150 MHz, 303 K, $CDCl_3$): δ 173.2, 133.6, 127.9, 110.0, 79.0, 74.6, 72.4, 35.7, 35.0, 28.2, 27.0, 25.8, 25.7, 17.9, 14.1. MS m/z (rel. intensity): 268 ([M^+], 23), 253 (10), 183 (24), 165 (12), 139 (19), 138 (30), 126 (22), 125 (33), 123 (12), 110 (16), 109 (13), 98 (11), 97 (100).

Acknowledgement. Generous financial support of our programs by the Deutsche Forschungsgemeinschaft (Leibniz award program), the Fonds der Chemischen Industrie, and the ACS (Cope Scholar Funds) is gratefully acknowledged.

References

1. Reviews: (a) Trnka, T. M.; Grubbs, R. H. *Acc. Chem. Res.* **2001**, *34*, 18-29. (b) Fürstner, A. *Angew. Chem., Int. Ed.* **2000**, *39*, 3012-3043. (c) Grubbs, R. H.; Chang, S. *Tetrahedron* **1998**, *54*, 4413-4450. (d) Schrock, R. R.; Hoveyda, A. *Angew. Chem. Int. Ed.* **2003**, *42*, 4592-4633. (e) Connon, S. J.; Blechert, S. *Angew. Chem., Int. Ed.* **2003**, *42*, 1900-1923. (f) Fürstner, A. *Top. Catal.* **1997**, *4*, 285-299.
2. For reviews on applications in carbohydrate chemistry see: (a) Roy, R.; Das, S. K. *Chem. Commun.* **2000**, 519-529. (b) Jorgensen, M.; Hadwiger, P.; Madsen, R.; Stütz, A. E.; Wrodnigg, T. M. *Curr. Org. Chem.* **2000**, *4*, 565-588.
3. For other projects of this laboratory dealing with carbohydrate chemistry see *i.a.*: (a) Fürstner, A.; Albert, M.; Mlynarski, J.; Matheu, M.; DeClercq, E. *J. Am. Chem. Soc.* **2003**, *125*, 13132-13142. (b) Fürstner, A.; Albert, M.; Mlynarski, J.; Matheu, M. *J. Am. Chem. Soc.* **2002**, *124*, 1168-1169. (c) Fürstner, A.; Mlynarski, J.; Albert, M. *J. Am. Chem. Soc.* **2002**, *124*, 10274-10275. (d) Fürstner, A.; Radkowski, K.; Grabowski, J.; Wirtz, C.; Mynott, R. *J. Org. Chem.* **2000**, *65*, 8758-8762. (e) Fürstner, A.; Konetzki, I. *J. Org. Chem.* **1998**, *63*, 3072-3080. (f) Fürstner, A.; Konetzki, I. *Tetrahedron* **1996**, *52*, 15071-15078. (g) Fürstner, A.; Ruiz-Caro, J.; Prinz, H.; Waldmann, H. *J. Org. Chem.* **2004**, *69*, 459-467. (h) Mlynarski, J.; Ruiz-Caro, J.; Fürstner, A. *Chem. Eur. J.* **2004**, in press.
4. Review: Balci, M. *Pure Appl. Chem.* **1997**, *69*, 97-104.
5. Ackermann, L.; El Tom, D.; Fürstner, A. *Tetrahedron* **2000**, *56*, 2195-2202.
6. For other RCM approaches to conduritols see: (a) Jorgensen, M.; Iversen, E. H.; Paulsen, A. L.; Madsen, R. *J. Org. Chem.* **2001**, *66*, 4630-4634. (b) Heo, J.-N.; Holson, E. B.; Roush, W. R. *Org. Lett.* **2003**, *5*, 1697-1700. (c) Kadota, K.; Takeuchi, M.; Taniguchi, T.; Ogasawara, K. *Org. Lett.* **2001**, *3*, 1769-1772. (d) Conrad, R. M.; Grogan, M. J.; Bertozzi, C. R. *Org. Lett.* **2002**, *4*, 1359-1361. (e) Lee, W.-W.; Chang, S. *Tetrahedron: Asymmetry*

1999, *10*, 4473-4475. (f) Gallos, J. K.; Koftis, T. V.; Sarli, V. C.; Litinas, K. E. *J. Chem. Soc., Perkin Trans. 1* **1999**, 3075-3077.
7. (a) Nguyen, S. T.; Grubbs, R. H.; Ziller, J. W. *J. Am. Chem. Soc.* **1993**, *115*, 9858-9859. (b) Schwab, P.; Grubbs, R. H.; Ziller, J. W. *J. Am. Chem. Soc.* **1996**, *118*, 100-110.
8. Schrock, R. R.; Murdzek, J. S.; Bazan, G. C.; Robbins, J.; DiMare, M.; O'Regan, M. *J. Am. Chem. Soc.* **1990**, *112*, 3875-3886.
9. (a) Huang, J.; Stevens, E. D.; Nolan, S. P.; Petersen, J. L. *J. Am. Chem. Soc.* **1999**, *121*, 2674-2678. (b) Scholl, M.; Trnka, T. M.; Morgan, J. P.; Grubbs, R. H. *Tetrahedron Lett.* **1999**, *40*, 2247-2250. (c) Ackermann, L.; Fürstner, A.; Weskamp, T.; Kohl, F. J.; Herrmann, W. A. *Tetrahedron Lett.* **1999**, *40*, 4787-4790. (d) Fürstner, A.; Thiel, O. R.; Ackermann, L.; Schanz, H.-J.; Nolan, S. P. *J. Org. Chem.* **2000**, *65*, 2204-2207.
10. Rivero-Cruz, J. F.; García-Aguirre, G.; Cerda-García-Rojas, C. M.; Mata, R. *Tetrahedron* **2000**, *56*, 5337-5344.
11. (a) Evidente, A.; Lanzetta, R.; Capasso, R.; Vurro, M.; Bottalico, A. *Phytochemistry* **1993**, *34*, 999-1003. (b) Evidente, A.; Capasso, R.; Abouzeid, M. A.; Lanzetta, R.; Vurro, M.; Bottalico, A. *J. Nat. Prod.* **1993**, *54*, 1937-1943. (c) de Napoli, L.; Messere, A.; Palomba, D.; Piccialli, V.; Evidente A.; Piccialli, G. *J. Org. Chem.* **2000**, *65*, 3432-3442.
12. Fürstner, A.; Radkowski, K. *Chem. Commun.* **2001**, 671-672.
13. Fürstner, A.; Radkowski, K.; Wirtz, C.; Goddard, R.; Lehmann, C. W.; Mynott, R. *J. Am. Chem. Soc.* **2002**, *124*, 7061-7069.
14. For an indirect solution of this problem see: (a) Fürstner, A.; Guth, O.; Rumbo, A.; Seidel, G. *J. Am. Chem. Soc.* **1999**, *121*, 11108-11113. (b) Fürstner, A.; Mathes, C.; Lehmann, C. W. *Chem. Eur. J.* **2001**, *7*, 5299-5317.
15. Fürstner, A.; Guth, O.; Düffels, A.; Seidel, G.; Liebl, M.; Gabor, B.; Mynott, R. *Chem. Eur. J.* **2001**, *7*, 4811-4820.
16. Lee, C. W.; Grubbs, R. H. *Org. Lett.* **2000**, *2*, 2145-2147.
17. Fürstner, A.; Thiel, O. R.; Kindler, N.; Bartkowska, B. *J. Org. Chem.* **2000**, *65*, 7990-7995.
18. Díez, E.; Dixon, D. J.; Ley, S. V.; Polara, A.; Rodríguez, F. *Helv. Chim. Acta* **2003**, *86*, 3717-3729.
19. For follow up papers illustrating the concept of *E,Z*-control by catalyst tuning see: (a) Murga, J.; Falomir, E.; García-Fortanet, J.; Carda, M.; Marco, J. A. *Org. Lett.* **2002**, *4*, 3447-3449. (b) Gurjar, M. K.; Nagaprasad, R.; Ramana, C. V. *Tetrahedron Lett.* **2003**, *44*, 2873-2875. (c) Beumer, R.; Bayón, P.; Bugada, P.; Ducki, S.; Mongelli, N.; Sirtori, F. R.; Telser, J.; Gennari, C. *Tetrahedron Lett.* **2003**, *44*, 681-684. (d) Liu, D.; Kozmin, S. A. *Org. Lett.* **2002**, *4*, 3005-3007.

20. (a) Lindquist, N.; Fenical, W. *Tetrahedron Lett.* **1989**, *30*, 2735-2738. (b) The structure of ascidiatrienolide had originally been mis-assigned; the correct structure was established by total synthesis, cf.: Congreve, M. S.; Holmes, A. B.; Hughes, A. B.; Looney, M. G. *J. Am. Chem. Soc.* **1993**, *115*, 5815-5816.
21. (a) Niwa, H.; Inagaki, H.; Yamada, K. *Tetrahedron Lett.* **1991**, *32*, 5127-5128. (b) Niwa, H.; Watanabe, M.; Inagaki, H.; Yamada, K. *Tetrahedron* **1994**, *50*, 7385-7400.
22. Fürstner, A.; Schlede, M. *Adv. Synth. Catal.* **2002**, *344*, 657-665.
23. Short reviews: (a) Fürstner, A. *Eur. J. Org. Chem.* **2004**, 943-958. (b) Pereda-Miranda, R.; Bah, M. *Curr. Top. Med. Chem.* **2003**, *3*, 111-131.
24. (a) Pereda-Miranda, R.; Mata, R.; Anaya, A. L.; Wickramaratne, D. B. M.; Pezzuto, J. M.; Kinghorn, A. D. *J. Nat. Prod.* **1993**, *56*, 571-582. (b) Bah, M.; Pereda-Miranda, R. *Tetrahedron* **1996**, *52*, 13063-13080. (c) Bah, M.; Pereda-Miranda, R. *Tetrahedron* **1997**, *53*, 9007-9022.
25. (a) Larson, D. P.; Heathcock, C. H. *J. Org. Chem.* **1997**, *62*, 8406-8418. (b) Lu, S.-F.; O'yang, Q.; Guo, Z.-W.; Yu, B.; Hui, Y.-Z. *J. Org. Chem.* **1997**, *62*, 8400-8405.
26. Fürstner, A.; Müller, T. *J. Am. Chem. Soc.* **1999**, *121*, 7814-7821.
27. (a) Fürstner, A.; Müller, T. *J. Org. Chem.* **1998**, *63*, 424-425. (b) Lehmann, C. W.; Fürstner, A.; Müller, T. *Z. Kristallogr.* **2000**, *215*, 114-117.
28. Ono, M.; Nakagawa, K.; Kawasaki, T.; Miyahara, K. *Chem. Pharm. Bull.* **1993**, *41*, 1925-1932.
29. Fürstner, A.; Jeanjean, F.; Razon, P. *Angew. Chem., Int. Ed.* **2002**, *41*, 2097-2101.
30. (a) Fürstner, A.; Jeanjean, F.; Razon, P.; Wirtz, C.; Mynott, R. *Chem. Eur. J.* **2003**, *9*, 307-319. (b) Fürstner, A.; Jeanjean, F.; Razon, P.; Wirtz, C.; Mynott, R. *Chem. Eur. J.* **2003**, *9*, 320-326.
31. (a) Schmidt, R. R. *Angew. Chem., Int. Ed. Engl.* **1986**, *25*, 212-235. (b) Schmidt, R. R.; Kinzy, W. *Adv. Carbohydr. Chem. Biochem.* **1994**, *50*, 21-123.
32. Review: Toshima, K.; Tatsuta, K. *Chem. Rev.* **1993**, *93*, 1503-1531.
33. Fürstner, A. *Synlett* **1999**, 1523-1533.

Chapter 2

Synthesis of Stable Carbohydrate Mimetics

Maarten H. D. Postema, Jared L. Piper, Lei Liu, Venu Komanduri, and Russell L. Betts

Department of Chemistry, 243 Chemistry, Wayne State University, 5101 Cass Avenue, Detroit, MI 48202

The synthesis of a variety of stable carbohydrate mimetics using a RCM approach is discussed. An esterification-ring-closing metathesis (RCM) approach has been utilized for the preparation of a variety of alkyl and aryl *C*-glycosides. The synthesis of a number of *C*-saccharides will also be addressed. Yield optimization studies and the synthesis of the substrates will also be discussed.

Introduction

The goal at the outset of our synthetic carbohydrate program was to develop a unified approach to the synthesis of stable carbohydrate mimetics. *C*-Glycosides, compounds in which the interglycosidic oxygen had been replaced by a carbon atom (Figure 1), are a well known class of stable carbohydrate mimics. Herein, we wish to present a unified and convergent method to gain access to alkyl and aryl β-*C*-glycosides as well as a variety of β-*C*-saccharides.

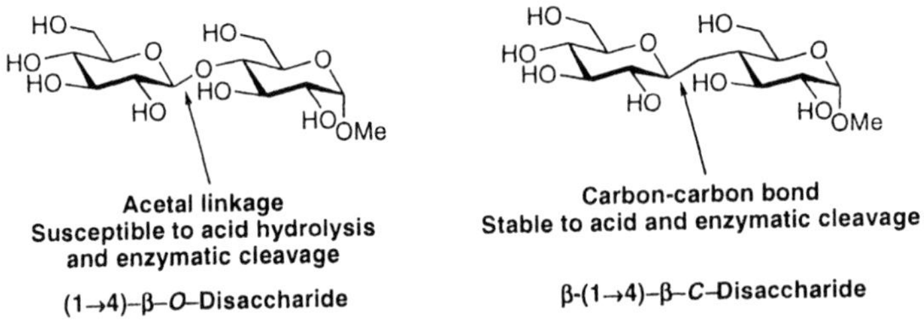

Acetal linkage
Susceptible to acid hydrolysis and enzymatic cleavage

(1→4)–β–*O*–Disaccharide

Carbon-carbon bond
Stable to acid and enzymatic cleavage

β-(1→4)–β–*C*–Disaccharide

Figure 1. Structures of O- and C-Glycosides

An esterification-RCM protocol is rather well-suited to fulfill this goal. The central scheme in Figure 2 illustrates the strategy. Esterification of olefin alcohol **1** with generic acid **2** should deliver ester **3**, which is subsequently converted to the glycal **4**, *via* a two-step protocol. Functionalization of the formed double bond of **4** delivers the target β-*C*-glycoside **5**. The convergence of this approach is obvious and the generality of access to various substrates is only limited by the availability of the required carboxylic acids. There have been several reviews written on the synthesis of *C*-glycosides (*1*). The author has made some contributions to this area as well (*2-5*). This chapter is a summary of an award lecture given by the author at the Fall 2003 ACS meeting held in New York City (*6*).

We consider the esterification reaction to be the cornerstone of our methodology. It is a convergent step that reliably produces the ester in a single and easy to perform step (Scheme 1).

This approach would neither be possible, nor conceivable, without the advent of modern olefin metathesis catalysts. Figure 3 shows a few of the most commonly used catalysts. In this work, we initially relied upon Schrock's Molybdenum catalyst **6** (*7*) to effect the ring closures, but now exclusively rely upon the second generation Grubbs ruthenium catalyst **7** (*8*).

The required olefin alcohols were prepared *via* a slight adaptation of the literature protocols (*9*). Kinetic furanoside formation was followed by benzylation (**10a** → **11a**) and hydrolysis produced the lactols **12a** that were subsequently purified by column chromatography. Wittig olefination then

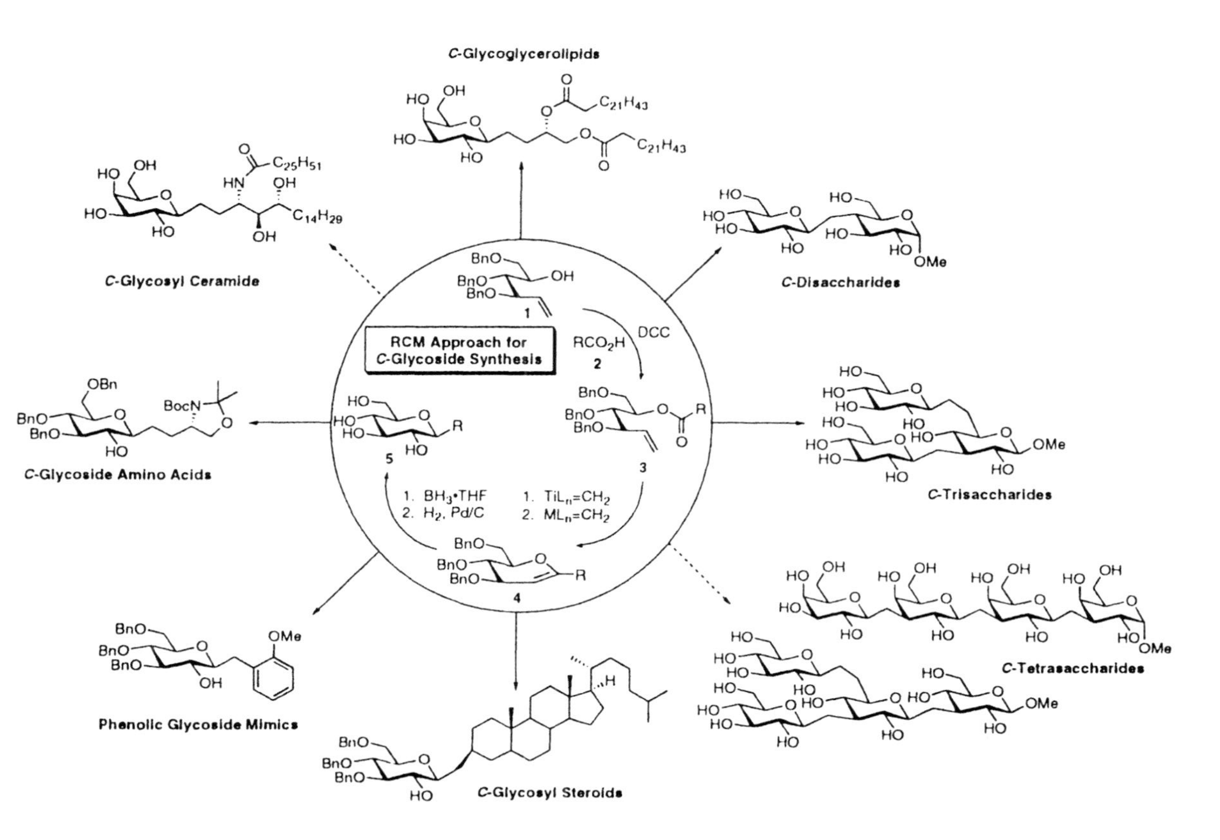

Figure 2. Our Program Plan

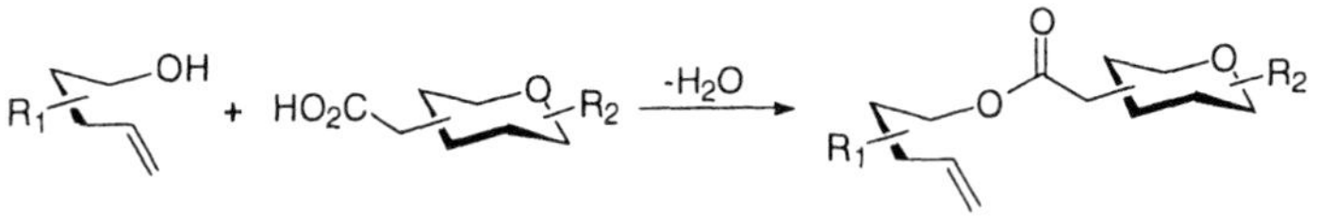

Scheme 1. *Esterification Reaction*

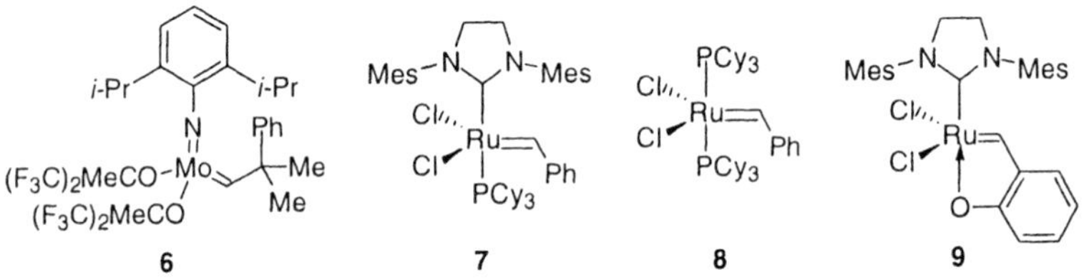

Figure 3. *RCM Catalysts*

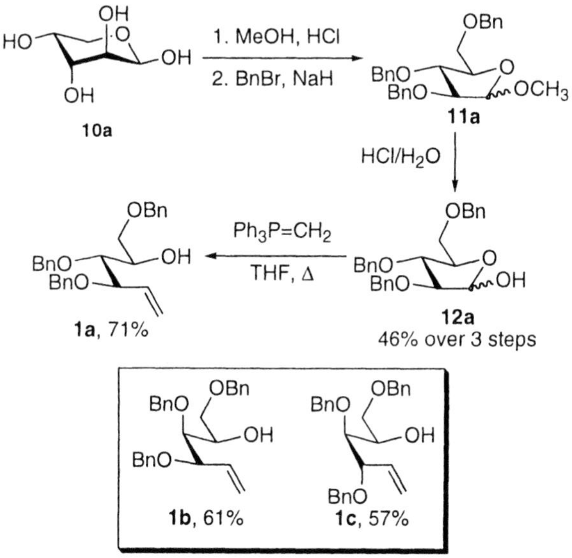

Scheme 2. Olefin Alcohol Synthesis

afforded the target alkenols **1a** (Scheme 2). Olefin alcohols **1b** and **1c** were prepared in a similar fashion by starting with the approriate pentose sugar.

Access to derivatives with protecting groups other than benzyl was also desired. Tri-*O*-acetyl-D-glucal (**13**) was deacetylated and benzylated with *p*-methoxybenzyl chloride to give **14**. Cleavage of the olefin in **14** gave formate aldehyde **15**. Hydrolysis of the formate ester led to lactol **16** and Wittig olefination then furnished the PMB protected olefin alcohol **1d** (Scheme 3).

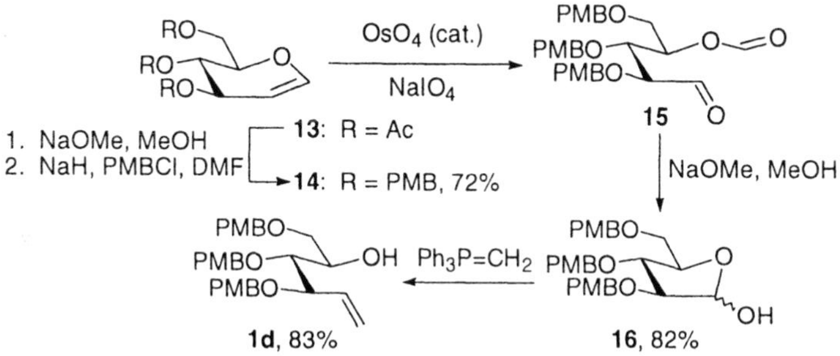

*Scheme 3. Synthesis of PMB Protected Olefin Alcohol **1d***

Synthesis of β-*C*-Glycosides

For the first example, we chose to acylate olefin alcohol **1a**. This was readily accomplished using acetic anhydride and 4-DMAP in pyridine to provide ester **17**. Methylenation, using Takai's (*10*) protocol, yielded the acyclic enol ether **18** which was subsequently cyclized with 15 mol % of the Schrock catalyst **6** in hot toluene to afford the glycal **19** in good yield. Hydroboration and oxidative work-up led to the methyl-*C*-glycoside **20** (Scheme 4). With this proof of principle in hand, we then set out to prepare a number of additional examples as shown in Table 1 (*11*).

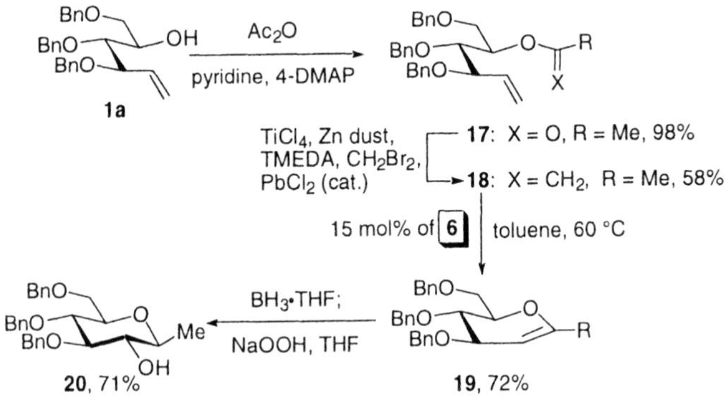

Scheme 4. Initial RCM Results

Table 1 shows the examples that were examined. Each example was prepared using a three-step protocol that includes DCC-mediated esterification, Takai methylenation and RCM with the Schrock catalyst **6**. The second entry underscores the need for use of a glovebox. In this case, the reaction was set-up in the glove-box and then refluxed under argon in the fumehood. The drop in yield, though not very large, is significant. We were unable to methylenate the *t*-butyl example (entry 3), although the cyclohexyl derivative (entry 4) behaved well. The styryl derivative (entry 5) did not undergo cyclization under the reaction conditions. To achieve a reasonable degree of cyclization with the aryl examples (entries 6-8), the amount of catalyst **6** had to be increased to 40 mol%.

Table 1. Synthesis of Alkyl and Aryl *C*-Glycosides

Entry	Ester/Acyclic Enol Ether	C-1 Glycal, (% Yield)[a, b]	Entry	Ester/Acyclic Enol Ether	C-1 Glycal, (% Yield)[a, b]
	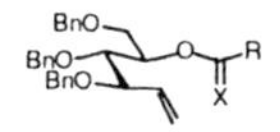			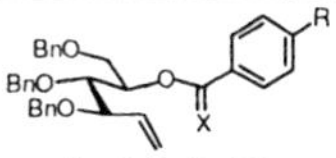	
1	X = O, R = Me, 98%[c] X = CH_2, R = Me, 58%	R = Me, 72%[d]	6	X = O, R = H, 68% X = CH_2, R = H, 58%	R = H, 57%[d]
2	X = O, R = *n*-Bu, 67% X = CH_2, R = *n*-Bu, 67%	R = *n*-Bu, 58%[e]	7	X = O, R = Br, 72% X = CH_2, R = Br, 64%	R = Br, 60%[d]
3	X = O, R = *t*-Bu, 42% X = CH_2, R = *t*-Bu, 0%	–[f]	8	X = O, R = OMe, 92% X = CH_2, R = OMe, 67%	R = OMe, 55%[d]
	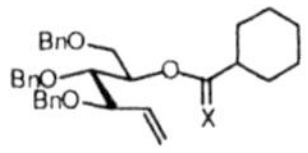			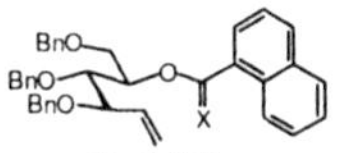	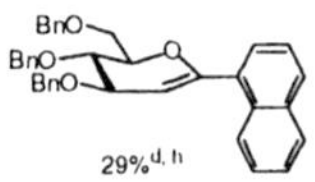
4	X = O, 68% X = CH_2, 61%	73%[d]	9	X = O, 79% X = CH_2, 64%	29%[d, h]
	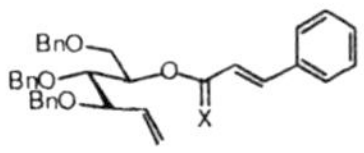			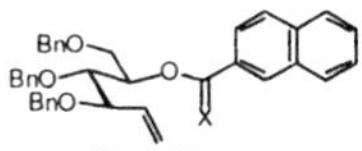	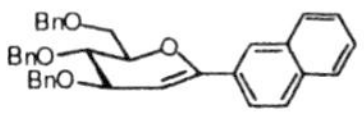
5	X = O, 79% X = CH_2, 62%	–[g]	10	X = O, 79% X = CH_2, 54%	68%

[a]Yields refer to chromatographically homogeneous (^{1}H NMR, 400 MHz) material. [b]Reaction was carried out on 40-70 mg scale at 0.01-0.02 M in substrate using 15 mol % **6**. [c]Reaction was carried out using Ac_2O, 4-DMAP. [d]Reaction was carried out in the glove box. [e]Reaction was carried out under argon (rubber septum) on the benchtop. [f]Reaction was not carried out. [g]Only starting material was recovered (~95%). [h]28% Recovered starting material was isolated along with a second product (~30 %).

Synthesis of β-*C*-Disaccharides

It was reasoned that *C*-saccharides, specifically *C*-disaccharides should be accessible if an acid function could be selectively attached onto a pre-existing pyranose ring. Scheme 5 shows the premise for this sequence.

We elected to first prepare a 1,6-linked-*C*-disaccharide (*12*) since it possessed two linking atoms and was a less sterically hindered substrate than the corresponding 1,4-, 1,3-, or 1,2-linked derivatives.

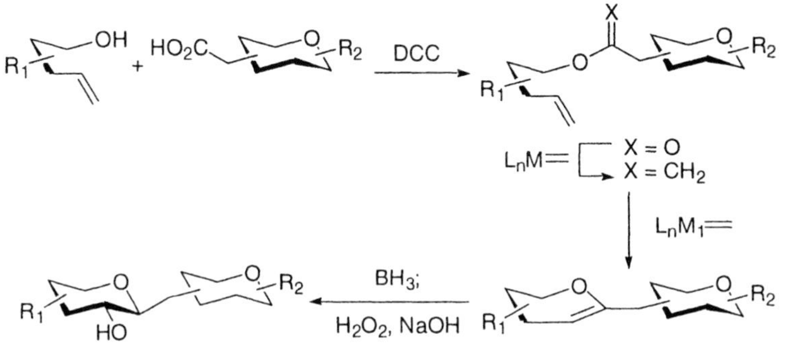

Scheme 5. A RCM Approach to β-C-Disaccharides

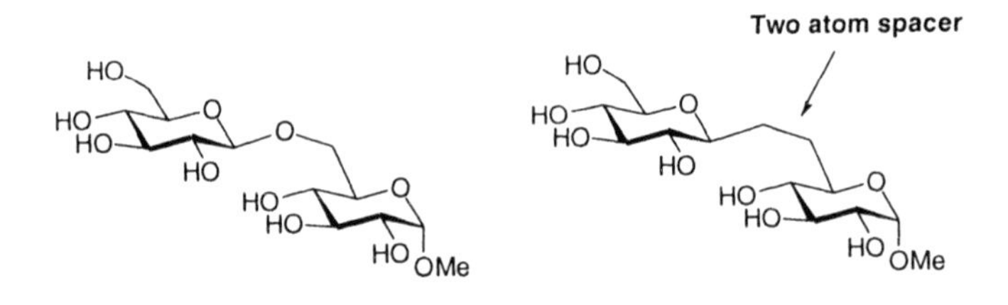

Figure 4. 1,6-Linked Disaccharides

The synthesis of the 1,6-linked ester was straightforward as outlined in Scheme 6. We were able to convert alcohol **23** to aldehyde **24** using Swern oxidation conditions. Wittig reaction was followed by olefin reduction and saponification to bring the sequence as far as **26**. Esterification of **26** with olefin alcohol **1a**, mediated by DCC, then afforded the target ester **27** in good overall yield (*13*).

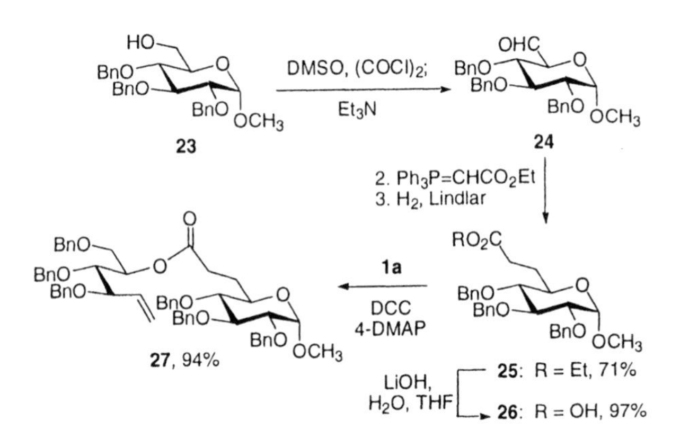

Scheme 6. Synthesis of the 1,6-Linked Ester

In a key experiment, it was demonstrated that methylenation of ester **27** gave **28** and RCM with 15 mol % of catalyst **6** gave a very fast reaction to produce the 1,6-glycal **29** in good yield (Scheme 7) (*13*). This was the most oxygenated example that had been cyclized with catalyst **6** at the time.

Next, the functionalization of **29** was examined and it was found that all the reactions gave good yield of addition and reduction products (Scheme 8).

A number of additional examples were studied. The carbohydrate-based carboxylic acids, the olefin alcohols and the protecting groups were all varied (Table 2) (*14*).

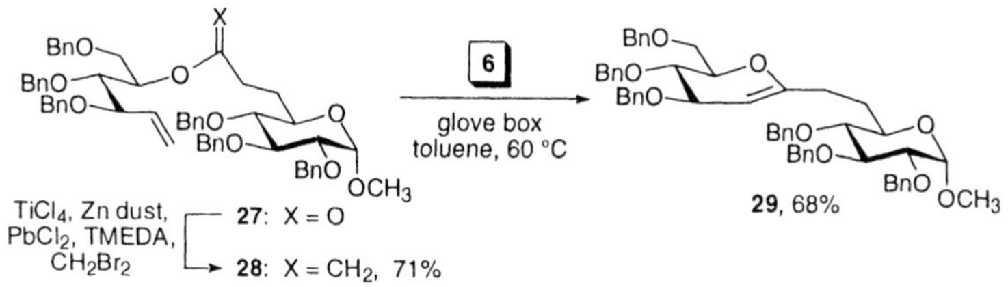

BnO BnO BnO BnO BnO BnO OCH3
TiCl4, Zn dust, PbCl2, TMEDA, CH2Br2
27: X = O
28: X = CH2, 71%
6
glove box
toluene, 60 °C
29, 68%

Scheme 7. Cyclization to the C-Disaccharidic Glycal

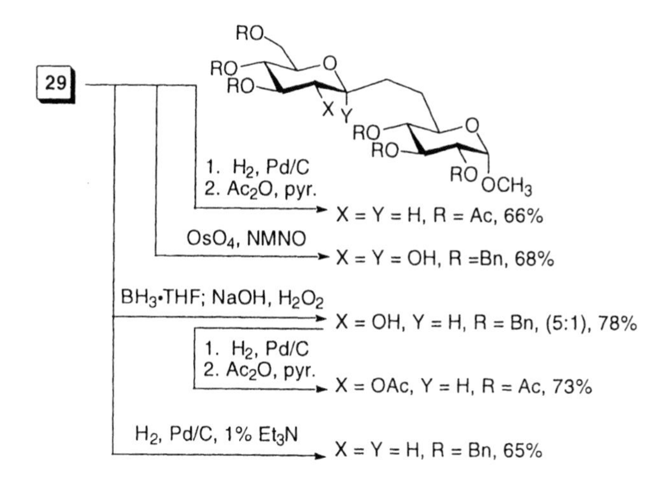

Scheme 8. Functionalization of the Olefin

Table 2. Synthesis of 1,6-Linked-*C*-Disaccharide Glycals

Entry	Ester/Acyclic Enol Ether	Glycal, (% Yield)[a, b]
1	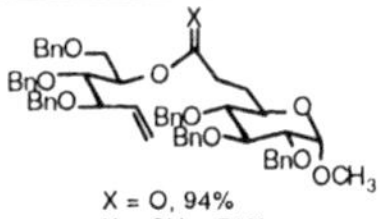 X = O, 94% X = CH_2, 71%	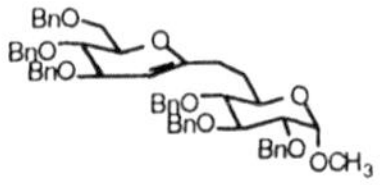 68%
2	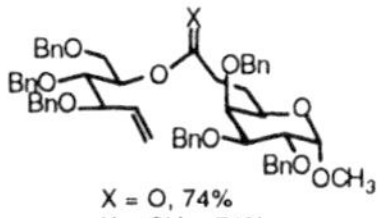 X = O, 74% X = CH_2, 71%	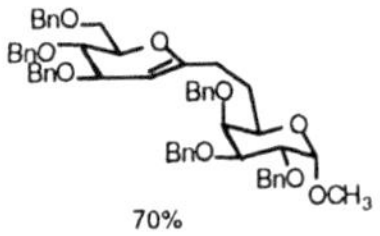 70%
3	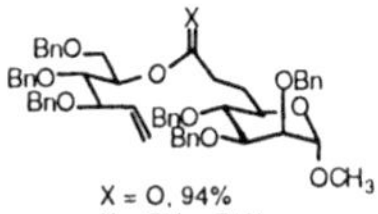 X = O, 94% X = CH_2, 71%	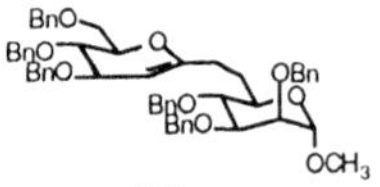 72%
4	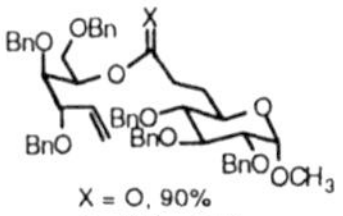 X = O, 90% X = CH_2, 71%	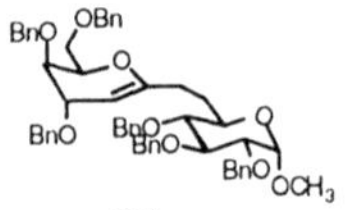 75%
5	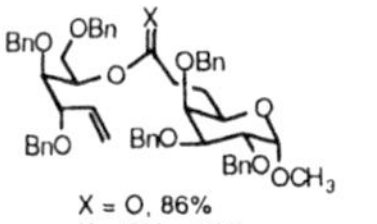 X = O, 86% X = CH_2, 67%	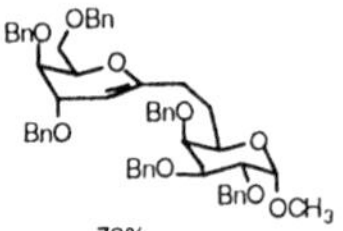 73%
6	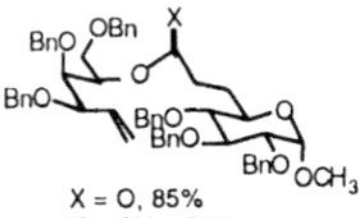 X = O, 85% X = CH_2, 70%	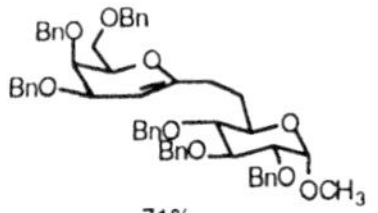 71%
7	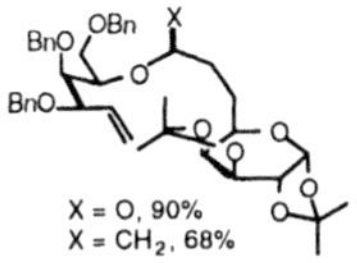 X = O, 90% X = CH_2, 68%	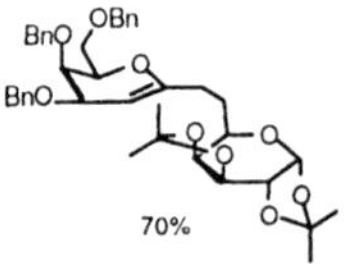 70%
8	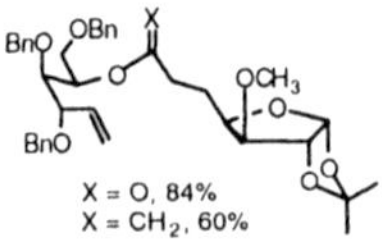 X = O, 84% X = CH_2, 60%	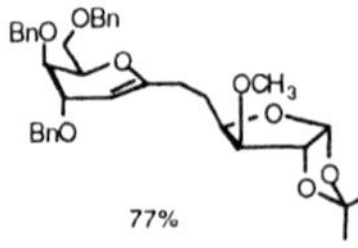 77%

[a]Yields refer to chromatographically homogeneous (^{1}H NMR, 500 MHz) material. [b]20-60 mg scale at 0.01-0.02 M in substrate using 20-30 mol % of metathesis catalyst **6**.

In an effort to make the approach more flexible, access to a small library of differentially-linked β-*C*-disaccharides (*15*) was considered. In this case, the acid functional group needed to be installed both in a regioselective and stereoselective fashion. This would then give access to a number of interesting *C*-disaccharide motifs (Figure 5).

After considerable experimentation, it was found that a Keck radical allylation approach (*16*) was a feasible method of installing the pendant allyl group. The radical allylation chemistry worked poorly when benzyl groups were present, presumably due to problems of abstraction of the benzylic hydrogens. This problem was easily circumvented by installing acetates as blocking groups. In this case, radical allylation furnished **32** as the major isomer (along with 11% of the axial isomer, not shown) with radical addition occurring *anti* to the two adjacent groups. The acetates were exchanged for benzyl groups (**32** → **33**) and only after cleavage of the olefin to the aldehyde were the epimers separable. Oxidation of the major equatorial aldehyde then afforded acid **34** in good yield (Scheme 9).

Ester formation proceeded smoothly to give **36**, however, cyclization of **36** using Shrock's catalyst gave a low yield of the product glycal **37** (Scheme 10). We reasoned that the glycal was unstable and perhaps decomposed during purification. This required us to adopt an alternate strategy that did not involve isolation of the glycal.

To avoid isolating the glycal, the reaction was quenched with an excess of borane in THF and, upon oxidative work-up under basic conditions, *C*-disaccharide **38** was obtained in 64% yield over two steps (Scheme 11) (*17*). The stereochemistry of the hydroboration step was verified on the corresponding acetate by examination of the $J_{1',2'}$ and $J_{2',3'}$ coupling constants in the proton NMR spectra.

At this point, the second generation Grubbs catalyst **7** (*8*) was reported and showed promise for use in such cyclizations. Using catalyst **7** and the one-pot protocol, *C*-disaccharide 4**3** was produced in 57% over two steps. It was also found that the stability of the formed glycals was not an issue since glycal **42** could be formed in 74% yield when catalyst **7** was used, compared to 40% (or less) when catalyst **6** was employed (Scheme 12) (*18*).

Heating of glycal **42** with 20 mol% of alcohol **44** (the ligand from the Shrock catalyst) resulted in the formation of the addition product **45** in about 20% yield along with some decomposition products (Scheme 13). Presumably, this is the cause of the lower yield when catalyst **6** is used for the cyclization. These results are preliminary and more studies need to be carried out to confirm this hypothesis.

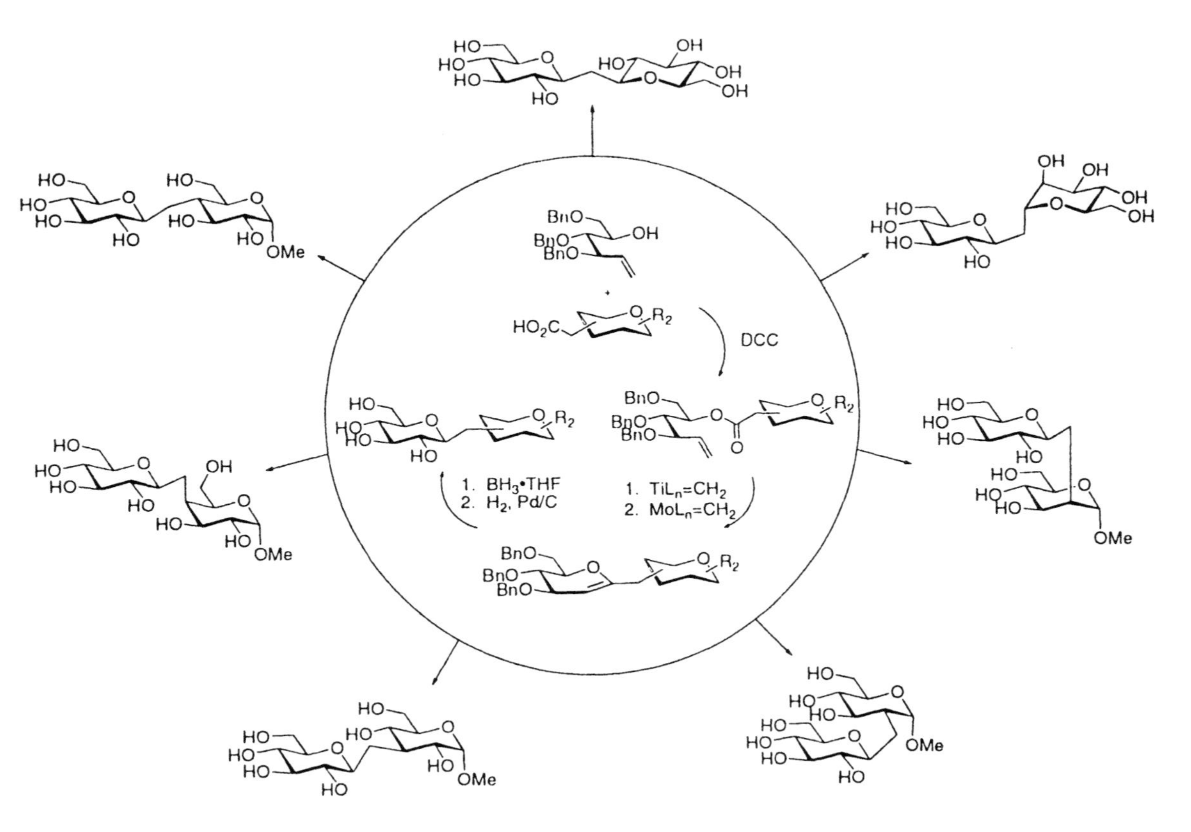

Figure 5. A Unified Approach to Differentially-Linked β-C-Disaccharides

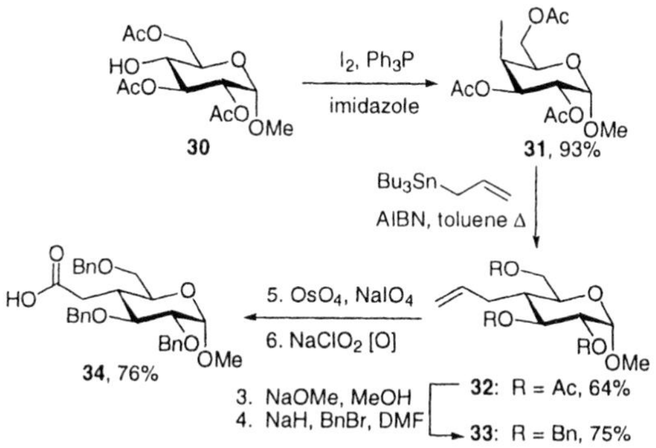

Scheme 9. *Keck Allylation Approach to the C-4 Acid*

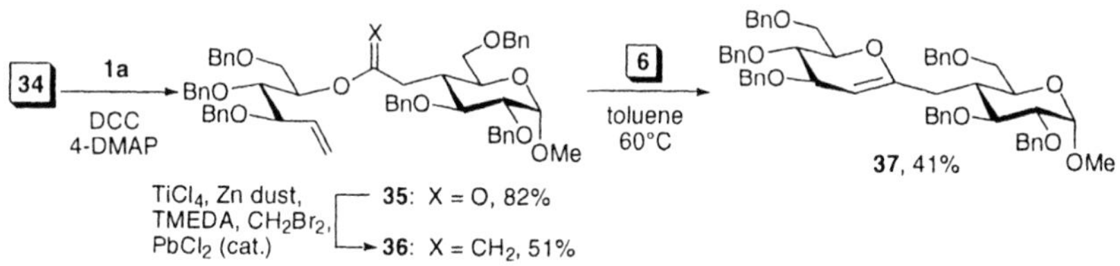

Scheme 10. *A Low Yielding Sequence*

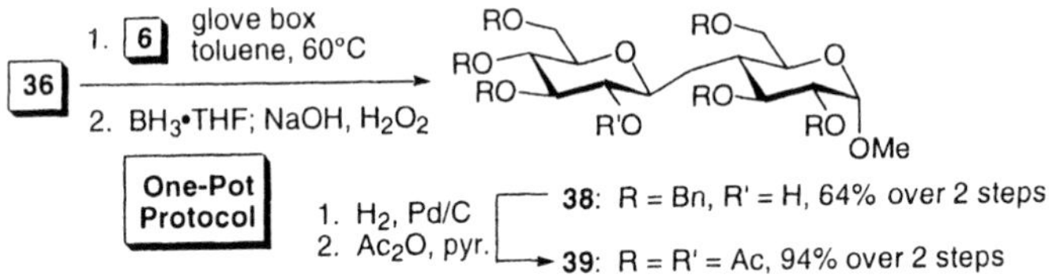

36 → 1. 6 glove box, toluene, 60°C; 2. BH_3•THF; NaOH, H_2O_2 (One-Pot Protocol)

1. H_2, Pd/C
2. Ac_2O, pyr.

38: R = Bn, R' = H, 64% over 2 steps

39: R = R' = Ac, 94% over 2 steps

Scheme 11. Implementation of the One-Pot Protocol

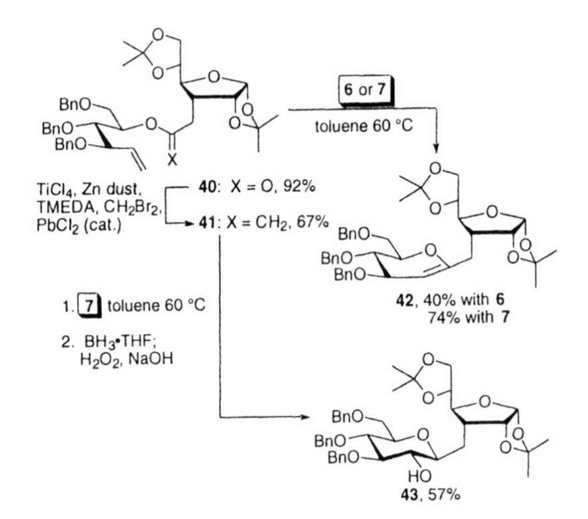

Scheme 12. Contrasting Results with Catalysts **6** *and* **7**

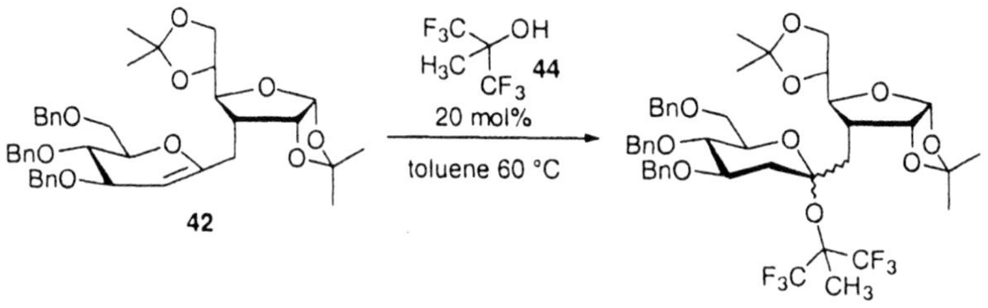

Scheme 13. A Possible Explanation

Cyclization of an example with no linking atoms demonstrates the power of this methodology to furnish direct linked or 1,1'-linked-*C*-disaccharides such as **47** (Scheme 14).

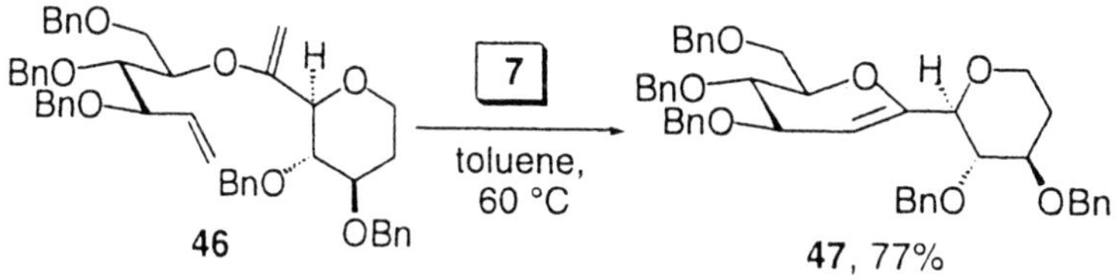

Scheme 14. Synthesis of the Direct-Linked Disaccharide ***47***

A variety of carbohydrate-based acids were then prepared *via* the free radical allylation route as shown below in Table 3 (*18*).

These acids were coupled to the appropriate olefinic alcohols and subsequent cyclization produced a small library of differentially-linked β-D-C-disaccharides (Table 4). The yields for the two-step procedure are good and entry 1 gives a side-by-side yield comparison of both catalysts. In essence, we were able to "walk around the ring" and install acid functionality at any carbon atom and selectively prepare the corresponding *C*-disaccharide (*18*).

Table 3. Synthesis of the Pyranose Acids

entry	alcohol	radical precursor	allyl derivatives[d,e]	acid[g,h]
1		85%	R = Ac, 64% R = Bn, 75%	76%
2			R = Ac, 11% R = Bn, 75%	77%
3	b	85%	R = Bz, 62% R = Bn, 94%	77%
4		85%	R = Ac, 26% R = Bn, 92%	78%
5			R = Ac, 38% R = Bn, 92%	84%
6	c	94%	R = Ac, 24%[f] R = Bn, 98%	60%

[a]Yields refer to chromatographically and spectroscopically homogeneous materials. [b]Prepared by the method of Szeja (*19*). [c]Prepared by the method of Hashimoto (*20*). [d]Any epimeric allylated derivatives were separated after benzylation and oxidative cleavage to the corresponding aldehydes. [e]Reaction was carried out with NaOMe in MeOH/THF followed by benzylation (BnBr/NaH/DMF). [f]The corresponding axial isomer was formed in 40% yield. [g]Oxidative cleavage of the olefin carried out in two steps. [h]Yields are for two steps.

Table 4. Synthesis of Differentially-Linked β-*C*-Disaccharides

Entry	Ester /Enol Ether	β-*C*-Disaccharide[d, e]
1	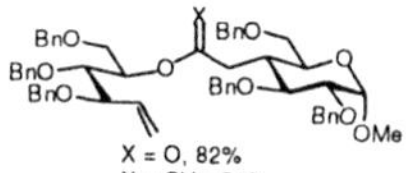 X = O, 82% X = CH_2, 51%	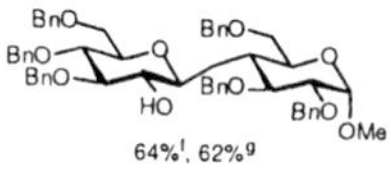 64%[f], 62%[g]
2	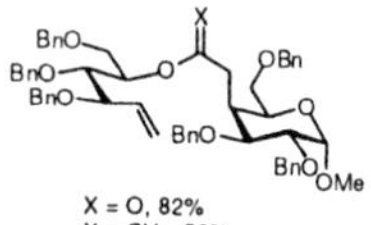 X = O, 82% X = CH_2, 52%	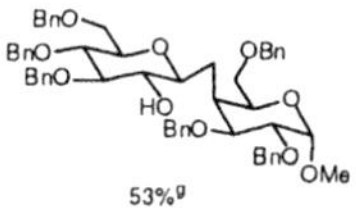 53%[g]
3	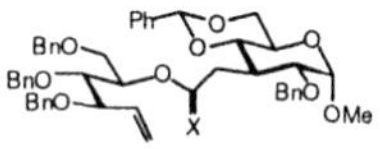 X = O, 92% X = CH_2, 58%	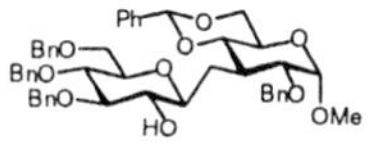 52%[g]
4	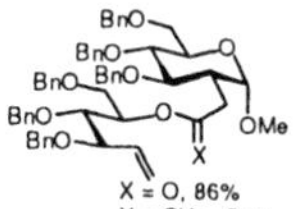 X = O, 86% X = CH_2, 51%	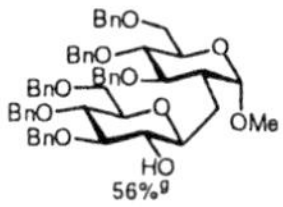 56%[g]
5	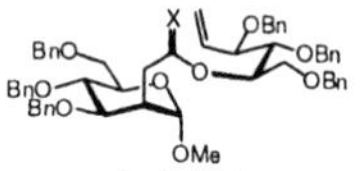 X = O, 88% X = CH_2, 52%	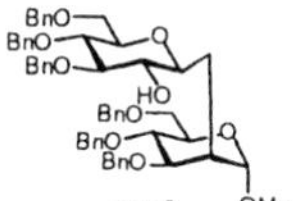 59%[g]
6	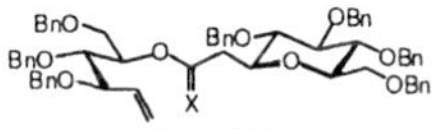 X = O, 77% X = CH_2, 50%	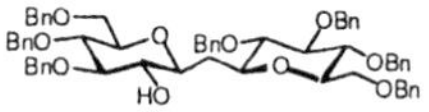 50%[g]
7	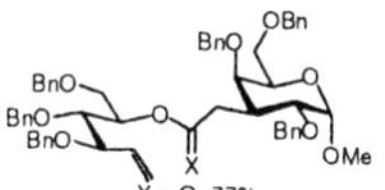 X = O, 77% X = CH_2, 50%	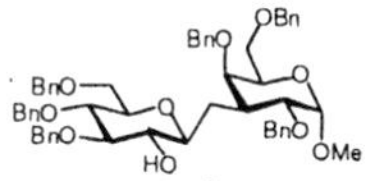 50%[g]
8	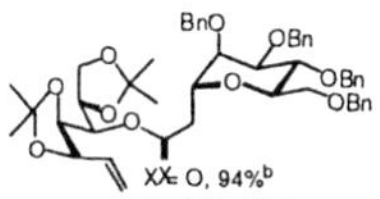 X = O, 94%[b] X = CH_2, 51%	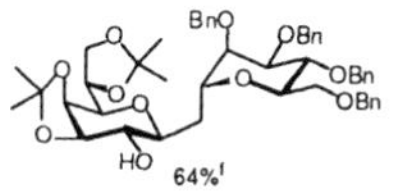 64%[f]
9	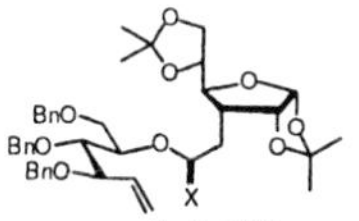 X = O, 92%[c] X = CH_2, 67%	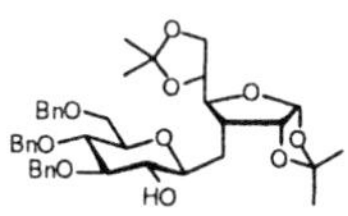 57%[g]

[a]Yields refer to chromatographically and spectroscopically homogeneous materials. [b]The corresponding olefin alcohol is known (*21*). [c]The corresponding ethyl ester is known (22). [d]Yields are for two steps; RCM and hydroboration-oxidative work-up. [e]Stereochemistry at *C*-1 and *C*-2 determined by acetylation and analysis of *H*-2 coupling constant in ^{1}H NMR. [f]Reaction carried out with 20-30 mol % of **6** in a glove box followed by hydroboration. [g]Reaction carried out with 20-30 mol % of **7** on an argon manifold followed by hydroboration.

Optimization of this reaction sequence was carried out on ester **27** as shown in Scheme 15.

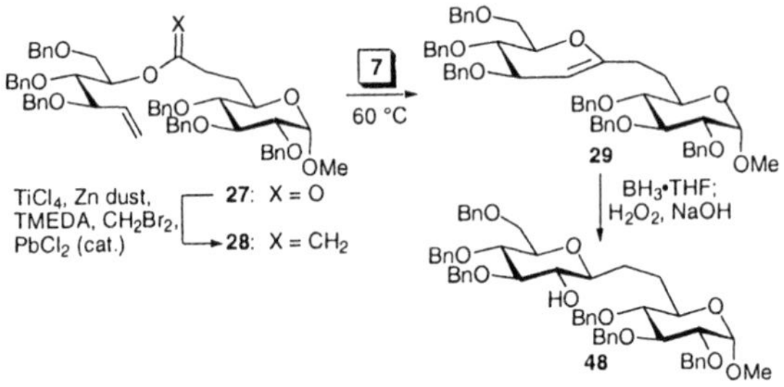

Scheme 15. Yield Optimization with Ester 27

The reaction was carried out in a few different ways as summarized below in Table 5. We found that material was being lost at the methylenation stage and that purification of the acyclic enol ether was causing a drop in the overall yield. If no purification was carried out until after the hydroboration step, then good overall yields (over 3 steps) of product could be obtained (Table 5) (*18*). To obtain reproducible yields, each of the individual reactions had to be pushed to completion.

Table 5. Yield Optimization

Yield of **28**	Yield of **29**	Yield of **48**	Overall Yield of **48**
76%[a]	68%	90%	47%
76%[a]	not purified[b]	65%[c]	49%
not purified[d]	58%[c]	90%	52%
not purified[d]	**not purified[e]**	**55%[f]**	**55%**

[a]The acyclic enol ether was purified by flash chromatography. [b]The product glycal was not isolated, but rather the one-pot protocol was employed. [c]Yield is over two steps. [d]The acyclic enol ether was not purified by flash chromatography, but the crude reaction mixture filtered through a pad of basic alumina. [e]The glycal was not isolated, but carried on crude to the next step. [f]Yield is over three steps.

The optimized protocol was then applied to a handful of substrates to furnish good yields (over 3 steps) of products (Table 6) (*18*).

Table 6. Yield Optimization of β-*C*-Disaccharide Synthesis

Entry	β-*C*-Disaccharide[b, c]	Peracetylated β-*C*-Disaccharide[e]
1	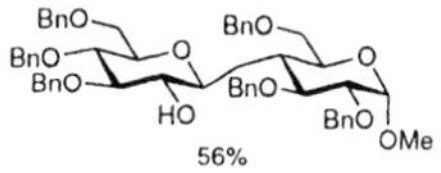 56%	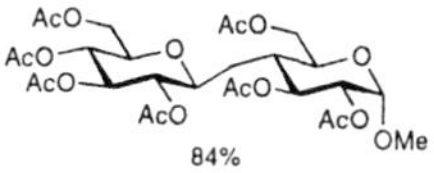 84%
2	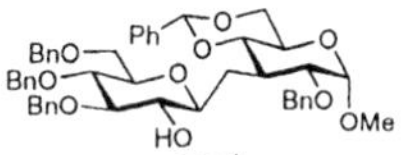 59%[d]	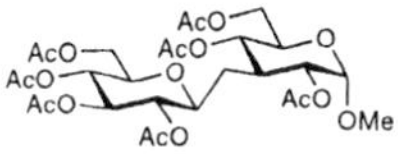 90%
3	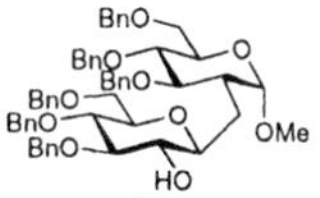 53%	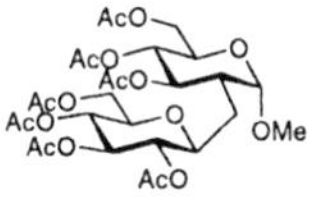 71%
4	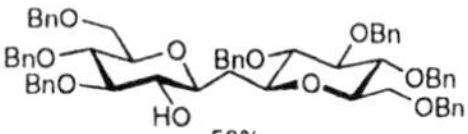 59%	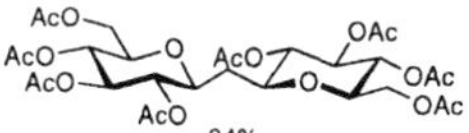 94%
5	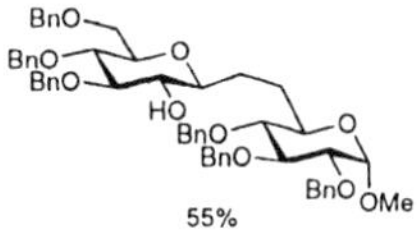 55%	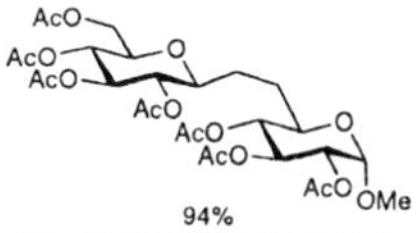 94%

[a]Yields refer to chromotographically and spectroscopically homogeneous material. [b]Yields are for three steps, methylenation, RCM and hydroboration/oxidative work-up. [c]RCM carried out with 20-25 mol% of **7**. [d]In this case, 40-45 mol% of **7** was required to push the reaction to completion. [e]Yields are for two steps, hydrogenation and peracetylation.

Synthesis of Biologically Relevant β-*C*-Glycoconjugates

This optimized protocol was then applied to the synthesis of a variety of potentially biologically significant *C*-glycosides. The acids were prepared *via* standard synthetic transformations and the yields for the three steps, methylenation, ring-closing metathesis and hydroboration were all over 50%. Table 7 shows the examples that were prepared (*23*). Entries 3 and 4 represent stable mimics of steroidal glycosides. Entries 5 and 6 are stable mimics of the corresponding *O*-linked amino acids and entries 7-9 are precursors to stable analogs of glyco*glycero*lipids (*24*).

Table 7. Synthesis of Biologically Relevant β-*C*-Glycosides

Entry	Ester/ Enol Ether	*C*-Glycoside[b,c]
1	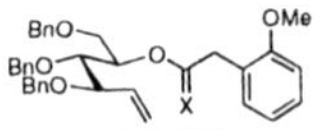 X = O, 86% X = CH_2	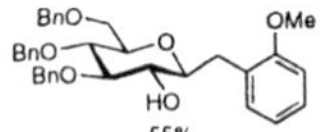 55%
2	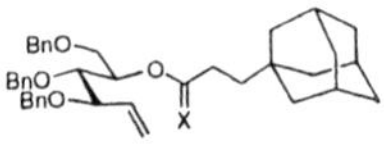 X = O, 93% X = CH_2	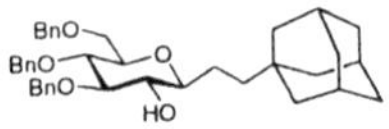 60%
3	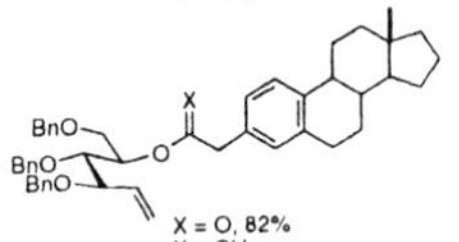 X = O, 82% X = CH_2	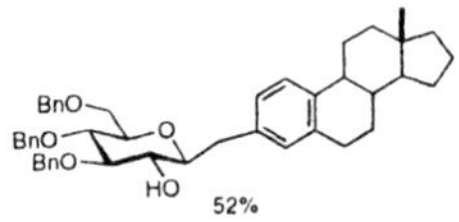 52%
4	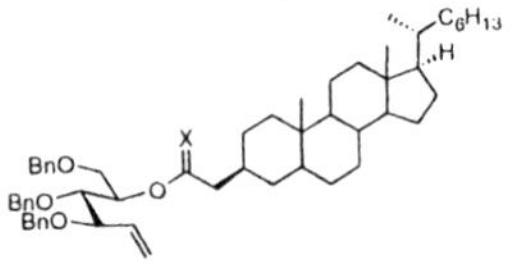 X = O, 91% X = CH_2	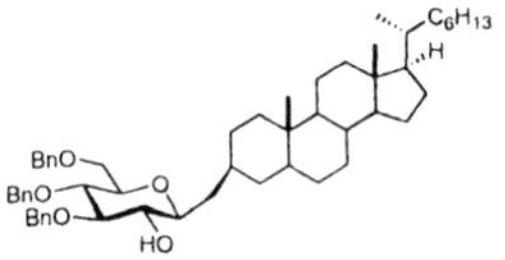 50%
5	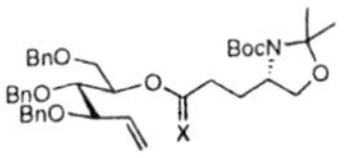 X = O, 85% X = CH_2	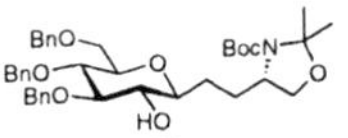 50%
6	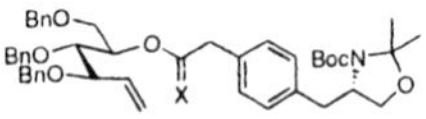 X = O, 82% X = CH_2	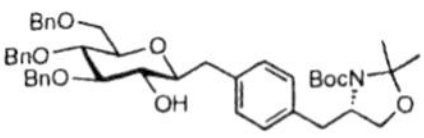 59%
7	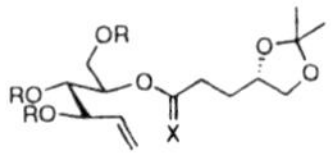 R = Bn, X = O, 94% R = Bn, X= CH_2	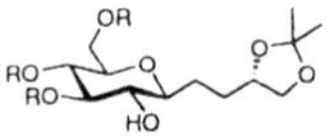 R = Bn, 57%
8	R = PMB, X = O, 87% R = PMB, X= CH_2	R = PMB, 50%
9	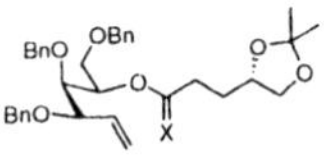 X = O, 97% X= CH_2	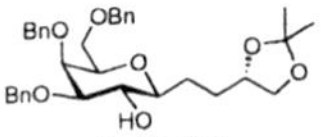 R = Bn, 57%

[a]All compounds have been fully characterized by standard spectral methods. [b]Formed by RCM with 20 mol% of **7** followed by hydroboration (BH_3•THF) and oxidative quench (NaOH, H_2O_2). [c]Yields are over three steps (methylenation, RCM, hydroboration and oxidative quench).

Scheme 17 shows the synthesis of a *C*-glucoglycerolipid. The secondary hydroxyl was capped (NaH, PMBBr) and the acetonide removed to deliver **51** (*23*). Diacylation then provided **52** which was deprotected to afford the target **53**. Selective acylation of the primary hydroxyl was readily accomplished by simply using one equivalent of the acid chloride.

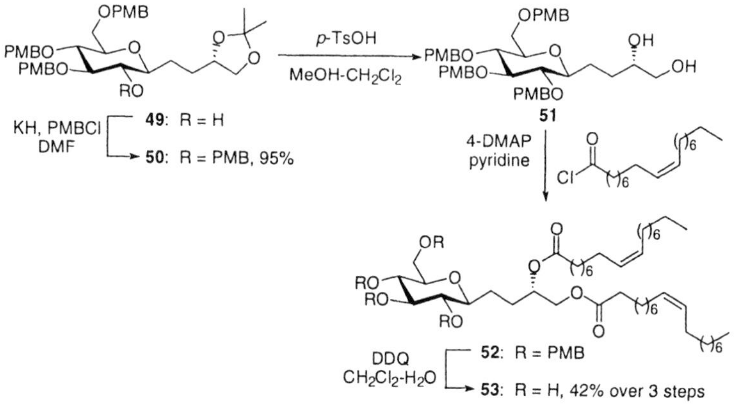

Scheme 17. Synthesis of a β-C-Glucoglycerolipid

A small library of these compounds was prepared and submitted for biological testing against a variety of solid tumor cell lines.

Synthesis of β-*C*-Trisaccharides

At this juncture, we felt confident that the chemistry was robust enough to explore a series of double cyclizations; in this way access to polyvalent structures would be possible. The application of our methodology to the synthesis of β-*C*-trisaccharides (*25*) was therefore considered (Scheme 18). In order to attempt the double cyclizations, it became clear that an efficient synthesis of the required diacid was required and that the possibility of macrocycle formation and other competing side reactions had to be considered.

Scheme 19 shows one synthetic approach to the needed diacids and illustrates the use of both Wittig and Keck allylation chemistry. Oxidation of the primary hydroxyl on **54** was followed by Wittig olefination and hydrogenation to reduce the formed alkene and remove the benzyl group. Installation of the radical precursor and subsequent free radical allylation gave a good yield of **56** as the sole isomer. Oxidative cleavage of the olefin and saponification of the ester provided diacid **57** in good yield, Scheme 19.

Methylenation of **58** with an excess of the Takai reagent and subsequent RCM with the second generation Grubbs catalyst **7** (35 mol%) gave an intermediate (bis)-*C*-glycal which was directly subjected to hydroboration by BH_3•THF and subsequent oxidative work-up to afford the *C*-trisaccharide **60** in 49% yield over three steps (*26*). Although double RCM reactions are known (*27*), this was the first time it has been used in the context of *C*-saccharide synthesis.

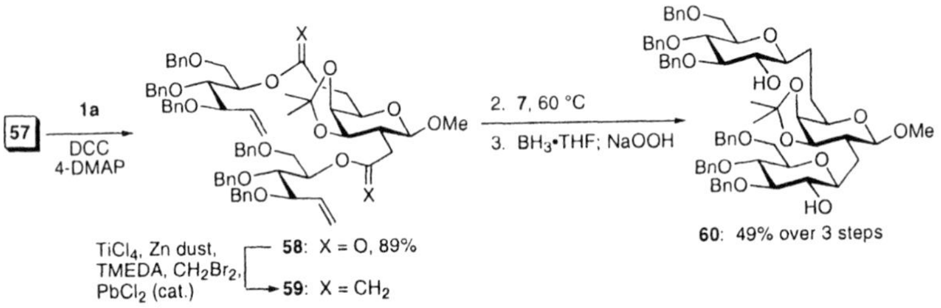

Scheme 20. Double RCM-Based Synthesis of β-C-Trisaccharides

Table 8 shows the examples that have been prepared thus far. Ester formation, mediated by DCC and 4-DMAP proved to be routine. A large excess of the methylenating reagent was required for the methylenation reactions to be driven to completion. In all the cases, 35 to 40 mol% of the RCM catalyst **7** was needed to achieve complete reaction. It is noteworthy that the three-step protocol works quite well even when both groups to be cyclized are on the same side of the pyranose ring as in entry 3. No evidence of other cyclized products by TLC or NMR were noted in the crude reaction mixtures.

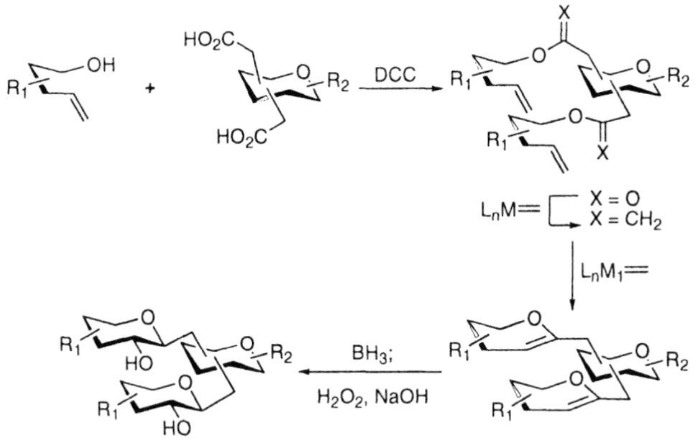

R$_1$ OH + HO_2C O R$_2$ HO_2C —DCC→ R$_1$ O X O R$_2$ R$_1$ O X

L_nM= X = O, X = CH_2

L_nM_1=

R$_1$ O HO O R$_2$ R$_1$ O HO ←BH$_3$; H_2O_2, NaOH— R$_1$ O O R$_2$ R$_1$ O

Scheme 18. Our Route to β-C-Trisaccharides

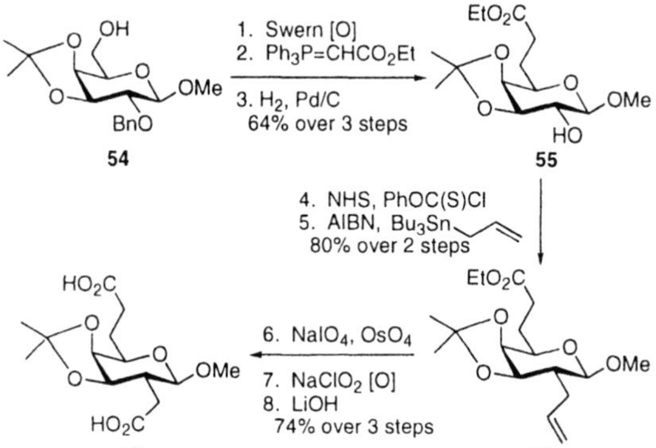

54: O OH O OMe O BnO

1. Swern [O]
2. Ph_3P=$CHCO_2Et$
3. H_2, Pd/C
64% over 3 steps

55: EtO_2C O O OMe O HO

4. NHS, PhOC(S)Cl
5. AIBN, Bu_3Sn
80% over 2 steps

56: EtO_2C O O OMe O

6. $NaIO_4$, OsO_4
7. $NaClO_2$ [O]
8. LiOH
74% over 3 steps

57: HO_2C O O OMe O HO_2C

Scheme 19. Synthesis of Diacid 57

Table 8. Synthesis of β-*C*-Trisaccharides by Double RCM[a,b]

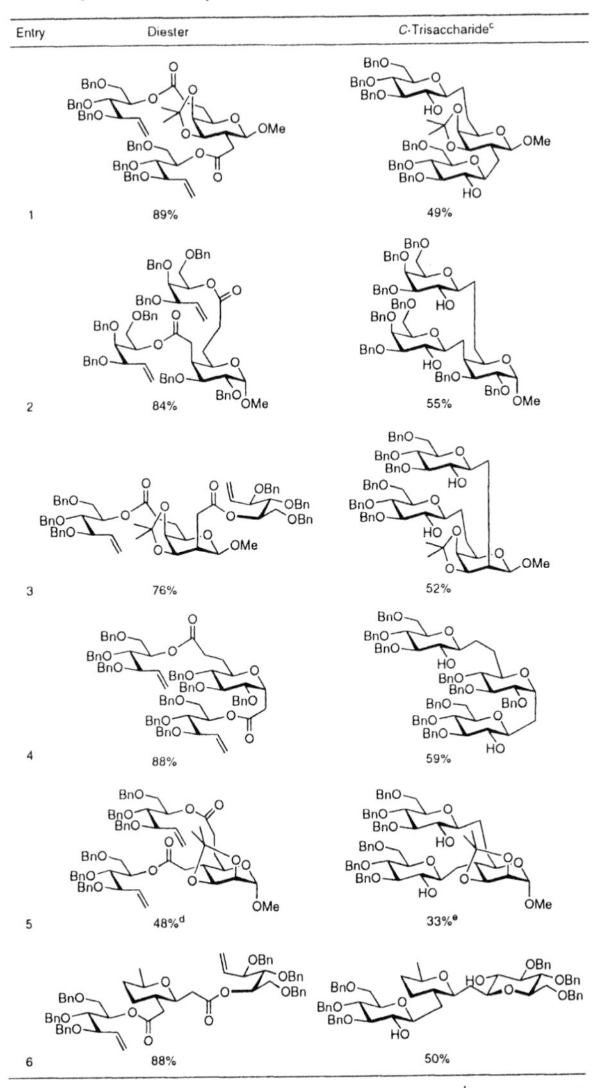

[a]Yields refer to chromatographically homogeneous material. [b]Yields are for three steps; methylenation, RCM (35-40 mol% of **7**) and hydroboration-oxidative work-up. [c]Stereochemistry at *C*-1 and *C*-2 determined by acetylation and analysis of *H*-2 coupling constant in ^{1}H NMR. [d]A fair amount of unreacted mono-1,6-ester was isolated from the reaction mixture. [e]In this case, the RCM reaction was stopped early, the (bis)*C*-glycal was isolated and purified (48%, unoptimized) and then subjected to hydroboration (66%, unoptimized).

Conclusion

The above results clearly demonstrate that our double RCM approach for *C*-trisaccharide synthesis is a viable, elegant and efficient approach to this important class of compounds and should provide the means to quickly assemble substrates for further biological or conformational studies. The synthesis of more complex and biologically relevant structures is underway and will be reported in due course. The author would like to express his appreciation and thanks to the co-workers whose names appear in this article and in the references. Without their hard work and dedication to the art, the contents of these pages would only be ideas.

Experimental

Synthesis of C-Trisaccharide 60

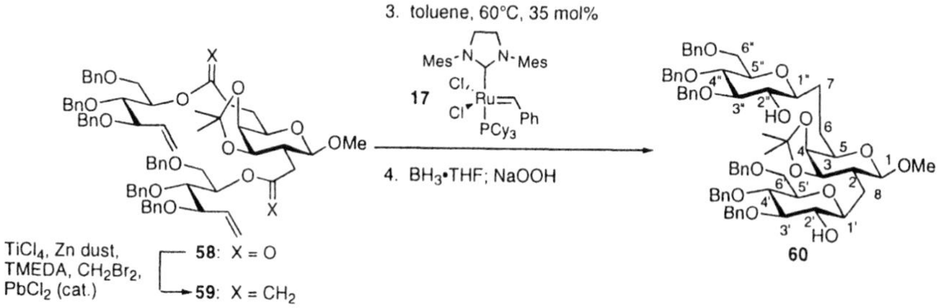

3. toluene, 60°C, 35 mol% **17**

4. BH_3•THF; NaOOH

$TiCl_4$, Zn dust, TMEDA, CH_2Br_2, $PbCl_2$ (cat.)

58: X = O

59: X = CH_2

60

A solution of titanium tetrachloride (0.75 mL, 2 M in CH_2Cl_2, 1.50 mmol) was added to cool (0 °C) THF (3 mL). The resulting mixture was stirred for 30 minutes at which point TMEDA (0.44 mL, 2.90 mmol) was added in one portion. The resulting yellow-brown suspension was allowed to warm to ambient temperature and stirred for 30 minutes. At this point, zinc dust (215 mg, 0.329 mmol) and lead (II) chloride (2.5 mg, 0.001 mmol) were added in portion and stirring at ambient temperature was continued for 10 min. A solution of ester **58** (100 mg, 0.089 mmol) and dibromomethane (0.58 mL, 0.83 mmol) in THF (1 mL plus 1 mL rinse) was then added *via* cannula to the reaction flask in one portion. The mixture was stirred at 60 °C for 1 hour, cooled to 0 °C and then quenched by the addition of saturated potassium carbonate (1 mL). The resulting mixture was stirred for 30 minutes (while warming to ambient temperature) diluted with ether (20 mL) and stirred vigorously for 15 minutes. The resulting mixture was filtered through neutral alumina using a 3% triethylamine-ether as the eluent. The greenish-blue precipitate that resulted was crushed (mortar and pestle) and thoroughly extracted by vigorous stirring over diethyl ether (15-20 mL) for 30 minutes. The combined ethereal extracts were then concentrated *in vacuo* to give the acyclic enol ether **59**.

Ruthenium-based metathesis catalyst **17** (26 mg, 0.031 mmol, 35 mol%), was added in **5 portions** over 2 hours to a dry and degassed toluene (9.0 mL) solution of crude **59** from the above experiment and the resulting mixture (0.01 M in substrate) was heated to 60 °C for 3 h at which point TLC (silica, 30% EtOAc-hexanes) showed the reaction was complete. The resulting solution was concentrated and diluted with THF (5 mL) and cooled to 0 °C. BH_3•THF (0.89 mL, 1M in THF, 0.89 mmol) was added to the cooled solution and the resulting mixture was allowed to warm up to room temperature, with the ice-bath place and stirred overnight. The solution was cooled back down to 0 °C and NaOH (20 mL, 1 M, 20 mmol) and hydrogen peroxide (20 mL, 30% in water) were added in one portion. The solution was allowed to warm to room temperature over 2 hours. The solution was then extracted with EtOAc (3 x 10 mL) and the combined organic extracts were washed with saturated sodium thiosulfate (1 x 20 mL), brine (1 x 10 mL), dried over magnesium sulfate and filtered. The solution was concentrated and flash chromatography of the residue over silica using 30→50% Et_2O-hexanes gave **60** (48 mg, 49% over 3 steps) as a pure (R_f = 0.42, TLC silica, 40% EtOAc-hexanes; ^{1}H NMR, 500 MHz) white solid: m.p. = 39-42 °C; $[\alpha]_D$ = -11.9 (c = 1.0, $CHCl_3$); FT-IR (neat) 3446 (br), 2916, 2849, 1454, 1371, 1216, 1095 cm^{-1}; ^{1}H NMR (700 MHz, $CDCl_3$) δ ~367.24(m, 25 H, Ar*H*), ~227.18(m, 5 H, Ar*H*), 4.94 (d, 1 H, *J* = 11.9 Hz, OC*H*$_2$Ph), 4.93 (d, 1 H, *J* = 11.2 Hz, OC*H*$_2$Ph), 4.79 (d, 1 H, *J* = 11.2 Hz, OC*H*$_2$Ph), 4.78 (d, 1 H, *J* = 10.5 Hz, OC*H*$_2$Ph), 4.74 (d, 1 H, *J* = 11.9 Hz, OC*H*$_2$Ph), 4.73 (d, 1 H, *J* = 11.2 Hz, OC*H*$_2$Ph), 4.61 (d, 1 H, *J* = 12.6 Hz, OC*H*$_2$Ph), 4.61-4.58 (m, 2 H, 2 x OC*H*$_2$Ph), 4.58 (d, 1 H, *J* = 11.2 Hz, OC*H*$_2$Ph), 4.53 (d, 1 H, *J* = 11.9 Hz, OC*H*$_2$Ph), 4.46 (d, 1 H, *J* = 11.9 Hz, OC*H*$_2$Ph), 3.94 (dd, 1 H, *J* = 9.1, 4.9 Hz, *H*-3), 3.90 (d, 1 H, *J* = 9.1 Hz, *H*-1), 3.80 (dd, 1 H, *J* = 4.9, 1.4 Hz, *H*-4), 3.73-3.67 (m, 3 H, 3 x OC*H*), 3.67-3.60 (m, 3 H, *H*-4", 2 x OC*H*), 3.57-3.51 (m, 2 H, *H*-5, *H*-1'), 3.49-3.44 (m, 2 H, 2 x OC*H*), 3.42 (s, 3 H, OC*H*$_3$), 3.40-3.35 (m, 2 H, 2 x OC*H*), 3.32 (dd, 1 H, *J* = 9.1, 9.1 Hz, *H*-2"), 3.25 (dd, 1 H, *J* = 9.1, 9.1 Hz, OC*H*), 3.18 (ddd, 1 H, *J* = 9.1, 9.1, 2.1 Hz, *H*-1"), 2.11 (dddd, 1 H, *J* = 9.5, 9.5, 9.5, 3.0 Hz, *H*-7), 2.02 (dddd, 1 H, *J* = 9.8, 9.8, 9.8, 2.8 Hz, *H*-2), 1.94-1.87 (m, 2 H, 2 x *H*-6), 1.81 (ddd, 1 H, *J* = 11.2, 11.2, 2.8 Hz, *H*-8), 1.68 (ddd, 1 H, *J* = 10, 10, 10, 2.8 Hz, *H*-8), 1.52 (dddd, 1 H, *J* = 11.9, 11.9, 4.0 Hz, *H*-7), 1.41 (s, 3 H, (C*H*$_3$O)$_2$C), 1.26 (s, 3 H, (C*H*$_3$O)$_2$C); ^{13}C NMR (100 MHz, C_6D_6) δ 139.54, 139.47, 139.30, 139.14, 138.92, 138.94, 128.69, 128.67, 128.56, 128.51, 128.47, 128.30, 128.11, 127.93, 127.91, 127.68, 127.65, 127.61; 109.13, 104.11, 87.47, 87.18, 80.10,

79.96, 79.55, 79.51, 79.14, 78.87, 77.13, 75.44, 75.07, 75.01, 74.84, 74.74, 74.64, 74.63, 73.56, 73.52, 73.16, 69.53, 69.29, 55.76, 40.78, 32.99, 28.92, 28.52, 28.03, 26.91; HRMS (ES): calcd for $C_{66}H_{78}O_{14}Na$ $(M)^+$ 1117.5284, found 1117.5235.

Acknowledgements

We gratefully acknowledge the NSF (CHE-0132770) for support of this research. Acknowledgment is also made to the Donors of the Petroleum Research Fund, administered by the American Chemical Society for partial support of this research, (#33075-G1). Mr. Jared Piper gratefully held a David H. Green Memorial Fellowship and a Willard R. Lenz Memorial Fellowship. Dr. Lei Liu was the holder of a Wayne State Dissertation Award. We thank Dr. M. K. Ksebati (WSU) for his assistance over the years with NMR collection. We are also grateful to Professor Fred Valeriote for his continued support in running our samples in his *in vitro* solid tumor assay.

References

(1) For reviews on *C*-glycoside synthesis, see: a) Du, Y.; Linhardt, R. J.; Vlahov, I. R. *Tetrahedron* **1998**, *54*, 9913-9959. b) Beau, J.-M.; Gallagher, T. *Top. Curr. Chem.* **1997**, *187*, 1-54. c) Nicotra, F. *Topics Curr. Chem.* **1997**, *187*, 55-83. d) Togo, H.; He, W.; Waki, Y.; Yokoyama, M. *Synlett* **1998**, 700-717. e) Levy, D. E.; Tang, C. *"The Chemistry of C-Glycosides"*; First ed.; Elsevier Science: Oxford, 1995; Vol. 13. f) Herscovici, J.; Antonakis, K. In *Studies in Natural Product Chemistry, Stereoselective Synthesis*, A. U. Rahman, Ed.; Elsevier Science, 1992, Vol. 10, pp 337-403. g) Postema, M. H. D. *Tetrahedron* **1992**, *48*, 8545-8599.

(2) Postema, M. H. D. *Tetrahedron* **1992**, *48*, 8545-8599.

(3) Postema, M. H. D. *C-Glycoside Synthesis*; First ed.; CRC Press: Boca Raton, 1995.

(4) Postema, M. H., D.; Calimente, D. In *Glycochemistry: Principles, Synthesis and Applications*; Wang, P. G., Bertozzi, C., Eds.; Marcel Dekker: New York, 2000; pp 77-131.

(5) Liu, L.; McKee, M.; Postema, M. H. D. *Curr. Org. Chem.* **2001**, *5*, 1133-1167.

(6) The author would like to thank Professor Roy for the kind invitation to write this chapter and also for taking the time to edit the monograph.

(7) (a) Schrock, R. R.; Murdzek, J. S.; Bazan, G. C.; Robbins, J.; DiMare, M.; O'Regan, M. *J. Am. Chem. Soc.* **1990**, *112*, 3875-3886. (b) Feldman, J.;

Murdzek, J. S.; Davis, W. M.; Schrock, R. R. *Organometallics* **1989**, *8*, 2260-2265.
(8) Scholl, M.; Ding, S.; Lee, C. W.; Grubbs, R. H. *Org. Lett.* **1999**, *1*, 953-956.
(9) Our approach to this type of structure is based on previous work: a) Barker, R.; Fletcher, H. G. *J. Org. Chem.* **1961**, *26*, 4605-4609. b) Pearson, W. H.; Hines, J. V. *Tetrahedron Lett.* **1991**, *32*, 5513-5516. c) Freeman, F.; Robrage, K. D. *Carbohydr. Res.* **1987**, *171*, 1-11.
(10) Takai, K.; Kakiuchi, T.; Kataoka, Y.; Utimoto, K. *J. Org. Chem.* **1994**, *59*, 2668-2670.
(11) Postema, M. H. D.; Calimente, D. *J. Org. Chem.* **1999**, *64*, 1770-1771.
(12) For previous approaches to the synthesis of 1,6-linked-*C*-disaccharides see: a) Griffin, F. K.; Paterson, D. E.; Taylor, R. J. K. *Angew Chem. Int. Ed. Engl.* **1999**, *38*, 2939-2942. b) Leeuwenburgh, M. A.; Timmers, C. M.; van der Marel, G.; van Boom, J. H.; Mallet, J. M.; Sinaÿ, P. *Tetrahedron Lett.* **1997**, *38*, 6251-6254. c) Dondoni, A.; Zuurmond, H; Boscarato, A. *J. Org. Chem.* **1997**, *62*, 8114-8124. d) Kobertz, W. K.; Bertozzi, C. R.; Bednarski, M. D. *J. Org. Chem.* **1996**, *61*, 1894-1897. e) Armstrong, R. W.; Sutherlin, D. P. *Tetrahedron Lett.* **1994**, *35*, 7743-7746. f) Martin, O. R.; Lai, W. *J. Org. Chem.* **1993**, *58*, 176-185. g) Martin, O. R.; Xie, F.; Kakarla, R.; Benhamza, R. *Synlett* **1993**, 165-167. h) Baumberger, F.; Vasella, A. Helv. Chim. Acta. **1983**, *66*, 2210-2222. i) Rouzaud, D.; Sinaÿ, P. *J. Chem. Soc., Chem. Commun.* **1983**, 1353-1354.
(13) Postema, M. H. D.; Calimente, D. *Tetrahedron Lett.* **1999**, *40*, 4755-4759.
(14) Postema, M. H. D.; Calimente, D.; Liu, L.; Behrmann, T. L. *J. Org. Chem.* **2000**, *65*, 6061-6068.
(15) For some synthetic approaches to *C*-disaccharides, see: a) Postema, M. H. D.; Calimente, D.; Liu, L.; Behrmann, T. L. *J. Org. Chem.* **2000**, *65*, 6061-6068. b) Griffin, F. K.; Paterson, D. E.; Taylor, R. J. K. *Angew. Chem. Int. Ed. Engl.* **1999**, *38*, 2939-2942. c) Khan, N.; Cheng, X.; Mootoo, D. R. *J. Am. Chem. Soc.* **1999**, *121*, 4918-4919. d) Leeuwenburgh, M. A.; Timmers, C. M.; van der Marel, G.; van Boom, J. H.; Mallet, J. M.; Sinaÿ, P. *Tetrahedron Lett.* **1997**, *38*, 6251-6254. e) Dondoni, A.; Zuurmond, H; Boscarato, A. *J. Org. Chem.* **1997**, 62, 8114-8124. f) Mallet, A.; Mallet, J.-M.; Sinay, P. *Tetrahedron Asymm.* **1994**, 5, 2593-2608. g) Sutherlin, D. P.; Armstrong, R. W. *J. Org. Chem.* **1997**, *62*, 5267-5283. h) Martin, O. R.; Lai, W. *J. Org. Chem.* **1993**, *58*, 176-185.
(16) Keck, G. E.; Enholm, E. J.; Yates, J. B.; Wiley, M. R. *Tetrahedron* **1985**, *41*, 4079-4094.
(17) Liu, L.; Postema, M. H. D. *J. Am. Chem. Soc.* **2001**, *123*, 8602-8603.
(18) Postema, M. H. D.; Piper, J. L.; Liu, L.; Shen, J.; Faust, M.; Andreana, P. *J. Org. Chem.* **2003**, *68*, 4748-4754.
(19) Szeja, W. *Carbohydr. Res.* **1983**, *115*, 240-242.

(20) Kajihara, Y.; Kodama, H.; Endo, T.; Hashimoto, H. *Carbohydr. Res.* **1998**, *306*, 361-378.

(21) Haudrechy, A.; Sinaÿ, P. *J. Org. Chem.* **1992**, *57*, 4142-4151.

(22) Lourens, G. J.; Koekemoer, J. M. *Tetrahedron Lett.* **1975**, *16*, 3719-3722.

(23) Postema, M. H. D.; Piper, J. L. *Org. Lett.* **2003**, *5*, 1721-1723.

(24) For previous syntheses of such compounds, see: (a) Cipolla, L.; Nicotra, F.; Vismara, E.; Guerrini, M. *Tetrahedron* **1997**, *53*, 6163-6170. (b) Gurjar, M. K.; Reddy, R. *Carbohydr. Lett.* **1997**, *2*, 293-298. (c) Dodoni, A.; Perrone, D. Turturici, E. *J. Org. Chem.* **1999**, *64*, 5557-5564. (d) Yang, G.; Franck, R. W.; Byun, H.-S. Bittman, R.; Samadder, P.; Arthur, G. *Org. Lett.* **1999**, *1*, 2149-2151. (e) Yang, G.; Franck, R. W.; Bittman, R.; Samadder, P.; Arthur, G. *Org. Lett.* **2001**, *3*, 197-200.

(25) For some recent synthetic approaches to *C*-trisaccharides, see: a) Dondoni, A.; Marra, A. *Tetrahedron Lett.* **2003**, *44*, 4067-4071; b) Mikkelsen, L. M.; Krintel, S. L.; Jimenez-Barbero, J.; Skrydstrup, T. *J. Org. Chem.* **2002**, *67*, 6297-6308; c) Sutherlin, D. P.; Armstrong, R. W. *J. Org. Chem.* **1997**, *62*, 5267-5283; d) Sutherlin, D. P.; Armstrong, R. W. *J. Am. Chem. Soc.* **1996**, *118*, 9802-9803.

(26) Postema, M. H. D.; Piper, J. L.; Lei, L.; Komanduri, V. *Angew. Chem. Int. Ed. Engl.* **2004**, *43*, 1608-1615.

(27) For examples of double RCM reactions, see: a) Grubbs, R. H.; Fu, G. C. *J. Am. Chem. Soc.* **1992**, *114*, 7324-7325. b) Wallace, D. J.; Bulger, P. G.; Kennedy, D. J.; Ashwood, M. S.; Cottrell, I. F.; Dolling, U.-H. *Synlett* **2001**, *3*, 357-360. c) Wallace, D. J.; Cowden, C. J.; Kennedy, D. J. Ashwood, M. S.; Cottrell, I. F.; Dolling, U.-H. , *Tetrahedron Lett.* **2000**, *41*, 2027-2029. d) Schmidt, B.; Wildemann, H. *J. Chem. Soc., Perkin Trans. 1* **2000**, 2916-2925. e) Clark, J. S.; Hamelin, O. *Angew. Chem.* **2000**, *112*, 380-382; *Angew. Chem. Int. Ed. Engl.* **2000**, *39*, 372-374 and references cited therein.

Chapter 3

Synthesis of Neu5Ac, KDN, and KDO *C*-Glycosides

Chi-Chang Chen[1], Dino Ress[1], and Robert J. Linhardt[1,2,*]

[1]Departments of Chemistry, Medicinal and Natural Products Chemistry, and Chemical and Biochemical Engineering, University of Iowa, Iowa City, IA 52242
[2]Current address: Departments of Chemistry, Biology, and Chemical Engineering, Rensselaer Polytechnic Institute, Troy, NY 12180

Ulosonic acids are unique monosaccharides that are often found at the non-reducing terminus of glycoconjugates. Since glycoconjugates are typically catabolized by the stepwise enzymatic removal of their monosaccharide units, modification of the ulosonic acid component represents an intriguing target for blocking this transformation. Glycoconjugates containing a modified ulosonic acid would be expected to be resistant to catabolism and glycan maturation and thus, might have significant therapeutic potential. The possible applications of such derivatives would include: 1. inhibitors of enzymes, such as neuraminidases, acting on ulosonic acid containing molecules; 2. therapeutic glycoconjugate containing agents, resistant to catabolism and glycan maturation, having increased biological half-lives, such as stable analogs of the glycolipids GM4 and GM3; 3. as immunogens for the preparation of anti-carbohydrate, antibody-based therapeutics; or 4. as active vaccine agents in the treatment of diseases involving glycoconjugates, such as bacterial infections, viral infections and cancer. This review describes the design and synthesis of glycoconjugates containing ulosonic acid *C*-oligosaccharides.

Introduction

Ulosonic acids are a diverse family of unique, complex monosaccharides that serve many important biological functions (Figure 1) (*1-4*). The most common ulosonic acid, Neu5Ac is a constituent of many glycoconjugates, occupying the non-reducing end of oligosaccharide chains. Glycoconjugates containing Neu5Ac (and other ulosonic acids) mediate a number of important biological events. The most significant function may be associated with their negative charge. Many Neu5Ac residues are found in the glycoconjugates bound to human cells (*5*). This affords a charged shell covering cells that can prevent their aggregation by electrostatic repulsion and facilitate aggregation through calcium binding (*6*). The viscosity of many biological fluids is regulated by the introduction or release of Neu5Ac in the oligosaccharide portion of glycoprotein components (*7*). Neu5Ac also represents an important biological receptor domain (*1,2,8-14*). Cell surfaces containing this ligand interact with biomolecules such as receptors, hormones, enzymes, toxins and viruses. Cell-cell recognition between circulating leukocytes in blood vessels and endothelial cells is believed to occur, in part, through the interaction between mammalian lectins (selectins) and Neu5Ac containing oligosaccharide ligands (*9*). Neu5Ac, present in glycolipids such as GM4, interacts with adhesion molecules important in cell growth and tissue regeneration (*15*). Many pathogens also use Neu5Ac to localize on the surface of cells they infect (*16,17*). The capsular polysaccharide of a variety of pathogenic bacteria (*16,18,19*) as well as the *N*-linked oligosaccharide sequences of virus envelope glycoproteins, such as gp120 of the human immunodeficiency virus (HIV), contain high levels of Neu5Ac (*20*).

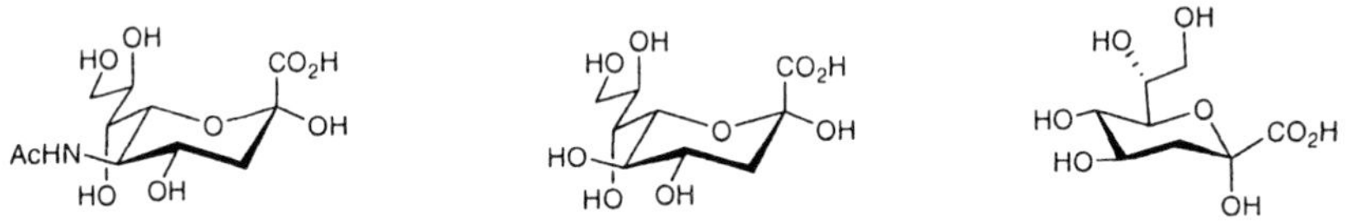

*Figure 1. The α-anomeric forms of ulosonic acids, Neu5Ac (**1**), KDN (**2**), and KDO (**3**).*

Other ulosonic acids, structurally related to Neu5Ac have been the focus of active scientific investigation due to their important biological functions. KDN (3-deoxy-D-*glycero*-D-*galacto*-2-nonulosonic acid) **2** bears a hydroxyl group in place of the acetamido group found at C-5 of Neu5Ac **1**. KDN is linked to galactosamine and galactose as the terminal unit of polysialoglycoproteins found in the membranes of mammalian tissues (*21*). KDO (3-deoxy-D-*glycero*-D-*manno*-2-octulosonic acid) **3** is a ketosidic component in all lipopolysaccharides (LPS) of gram-negative bacteria (*22*). It has also been identified in several acidic bacterial exopolysaccharides (K-antigens) (*23-25*). While the precise biological function of KDN and KDO is still undetermined, they also appear to play a role in cellular interactions with biomolecules.

The glycan component of glycoconjugates (*i.e.*, glycoproteins/peptides and glycolipids) are catabolized in both the extracellular environment and within cells. The turnover of endogenous glycoconjugates is initiated in the extracellular spaces where they are found, through the action of both exolytic and endolytic enzymes (*26-28*). Exo-glycosidases act sequentially to remove one saccharide unit at a time from the non-reducing end of a glycan chain. A fully elaborated biantennary glycan chain of a glycoprotein, for example must first be desialylated, through the action of a neuraminidase, prior to removal of the next residue, galactose. The resulting asialoglycoproteins have higher clearance rates due to the galactose-binding lectin in liver (*29,30*). In addition to their sequential breakdown by exolytic enzymes the glycan chains of a glycoconjugate can be removed through the action of endolytic glycosidases (*26-28*). For example, endoglycanases can remove intact glycans from glycoproteins/peptides and ceramidases can remove the intact glycans from glycolipids (*31*). Intracellular catabolism of glycoconjugates occurs in lysosomal compartments primarily through the sequential action of exolytic enzymes (*32-34*).

Neuraminidase action, important in most but not all glycoconjugate catabolism and maturation, represents a uniquely important therapeutic target. Neuraminidases, obtained from diverse species and tissues, interact with Neu5Ac present on the outer layer of cell membranes and result in its removal (*17*). This has a dramatic consequence for infection, adhesion and recognition events. Neuraminidases belong to a class of enzymes called hydrolases. Modified glycosides of Neu5Ac such as thioglycosides (*35-37*), initially synthesized to serve as powerful glycosylating reagents (*38*), are not neuraminidase substrates but instead act as inhibitors. Glycoconjugates terminated with other ulosonic acids such as KDN, are also catabolized through the action of hydrolases (*39*).

Recent developments in carbohydrate synthesis (*40*) opened up the possibility of forming the crucial *C*-glycosidic bond in Neu5Ac and other ulosonic acid containing oligosaccharides. Despite major advances in "*C*"-glycosylation chemistry (*40*), glycosylation reactions of Neu5Ac (and other ulosonic acids) proved particularly complex, often resulting in the 2,3-dehydro derivatives (*41*). These difficulties can be attributed to three factors inherent in the Neu5Ac molecule. First, the carboxyl group attached at C2 electronically disfavors oxocarbenium ion formation, an intermediate for almost all glycosidation reactions. Second, the carboxyl group sterically restricts glycoside formation. Third, the lack of a substituent at the adjacent C3 precludes assisting and directing effects.

Elegant methods for direct carbon-carbon (C-C) bond formation at the anomeric center in aldoses and ketoses (*42-45*), afforded only trivial alkyl and hydroxymethyl *C*-glycosides of Neu5Ac and other ulosonic acids (*46-49*). The anomeric outcome of the *C*-glycosylation may also be determined by the geometry of the enolate. The major problem confounding the synthesis of sialic acid *C*-glycosides appeared to be the requirement that the C-C bond being formed results in a new tertiary *C*-atom center. The driving force for developing new chemistry was driven by the importance of these targets. The biological significance of the *C*-glycosides of Neu5Ac and related ulosonic acids is obvious as they represent an important new class of hydrolytically stable analogs and mimetics of the natural *O*-glycosides. The use of Neu5Ac, KDN, and KDO *C*-glycosides (Figure 2) **4, 5,** and **6** as potential stable ligands for carbohydrate receptors may have a wide variety of potential therapeutic applications. This new class of compounds might play a role as glycoenzyme regulators. These "*C*"-glycosides might find use as potent inhibitors of neuraminidase (KDOase or KDNase) (*39*) and they might also be used as antiviral agents against neuraminidase-containing influenza (*50,51*) or as agents to block the removal of ulosonic acid containing glycoconjugates (*52*).

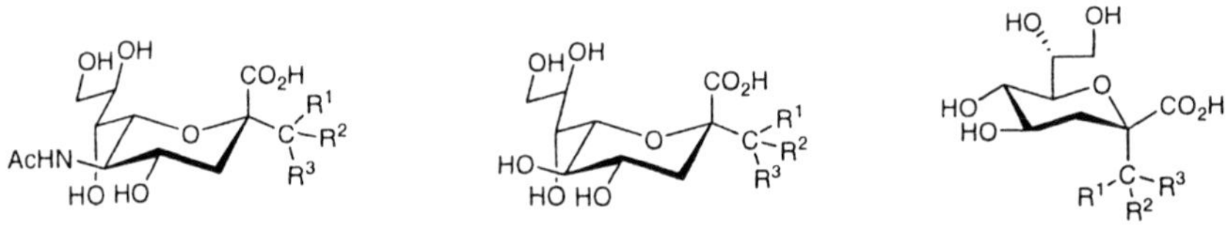

4 **5** **6**

Figure 2. Structure of α-"C"-glycosides of ulosonic acids, R^1=H or OH; R^2 = H or alkyl/aryl; R^3 = alkyl/ aryl/C-linked saccharide.

The half-life of many therapeutic glycoproteins, such as tissue plasminogen activator (tPA), is controlled by their neuraminidase catalyzed de-sialylation followed by their removal in the liver through galactose binding lectins (*53-55*). This clearance might be effectively blocked by inhibiting neuraminidase activity. *C*-glycosides might also be useful as receptor agonists/antagonists. The development of vaccines against carbohydrates is of crucial importance in the fields of therapeutic glycobiology and immunology (*56-57*). A significant portion of the anti-tumor response against cancers involves carbohydrate tumor antigens (*58*). Microorganisms also often express carbohydrate antigens and the immune response of the host to these antigens is an important mechanism of defense (*44*). Despite considerable interest, there remain many questions about host response to carbohydrate antigens. At present it is not known whether T-lymphocytes are able to recognize carbohydrate-defined epitopes and very little is understood about the fate of carbohydrates during antigen processing (*59*). It is also unclear why antigens show little T-cell dependent class switching from IgM to IgG and affinity maturation (*60*). It is clear that antibody affinity for carbohydrates can be significantly increased (*61*) through one or more of the following means: (1) the selection of potent adjuvants; (2) the use of highly antigenic carrier proteins (such as KLH) (*62,63*); and (3) increased carbohydrate valency (*62, 63*). Interestingly, the clustering of carbohydrate antigens on a carrier maximizes immunoreactivity (*62,63*). One approach for improving carbohydrate antigens that has not been thoroughly explored is vaccination over an extended length of time (*63*). It is here that a catabolically stable *C*-glycoside containing vaccine might play an important role. The increased conformational flexibility of ulosonic acid *C*-glycosides (*58, 64-69*) may affect their antigenicity (*70*) and also may result in altered binding affinity or cross-reactivity (*71*). Only the synthesis and testing of these *C*-glycoside targets as immunogens with flexible antigenic determinants can address these issues. Anti-carbohydrate, IgG-type antibodies might be used therapeutically in the application of targeting endogenous glycans, such as sLe^x containing ligands for the treatment of reperfusion injury (*72,73*). Alternatively, these agents might be useful in preparing immunogens for active immunization against ulosonic acid containing glycoconjugates in the design and preparation of anti-viral (*20, 74-77*), anti-bacterial (*16, 18, 23-25*) and anti-cancer vaccines (*56,57, 60, 62-64-78*).

Synthesis of Neu5Ac α-C-Glycosides 4 by Samarium-mediated Reduction

Samarium iodide, a powerful one electron reductant, has been used for C-C bond formation in carbohydrate synthesis by only a few research laboratories (*79, 80*). Encouraged by the reports of Sinaÿ (*79*) and Wong (*80*) on the samarium-mediated coupling of α-alkoxy sulfones with ketones, and glycosyl phosphates with ketones or aldehydes, we undertook a new approach for the synthesis of Neu5Ac α-*C*-glycosides relying on aryl sulfone donors (**7**) (Scheme 1). We initially selected Neur5Ac 2-pyridyl sulfone as our donor, minimizing the energy of one-electron transfer, to react with an aldehyde acceptor and afford the desired α-*C*-glycoside (*81*). Samarium mediated *C*-glycosylation was achieved in good yield with excellent stereoselectivity. Two competing mechanisms were advanced to provide a rational understanding of the mechanism of the SmI_2 mediated union of the sulfone donor with aldehyde and keto-based acceptors.

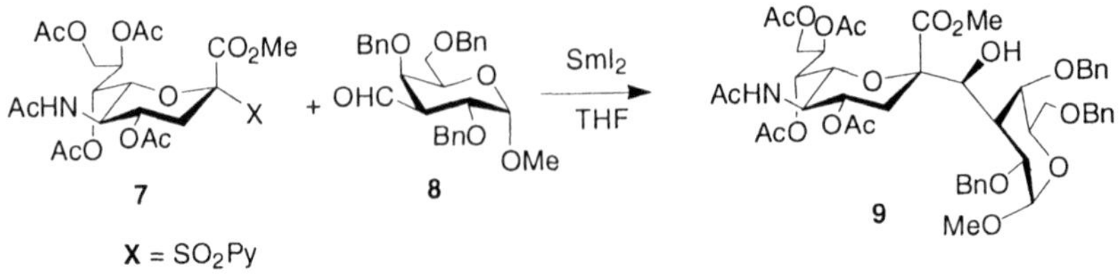

Scheme 1.

A simple radical mechanism can not explain the stereoselectivity of this reaction and an ionic mechanism does not take into account the propensity of samarium to form radical species. The stereoselective formation of *C*-glycoside is proposed to occur through two one electron transfer steps (**7** to **8**, **8** to **9**) yielding a samarium(III) glycoside intermediate (**9**) (Scheme 2). The resulting nucleophile is then believed to take advantage of the inherent oxophilic nature of the equatorial samarium to coordinate the stereoselective delivery of the incoming aldehyde or ketone electrophile to the bottom face (**10**). Hydrolysis of the samarium complexed product **11** affords *C*-glycoside **12**. This mechanism rationalizes α-stereoselectivity by invoking a *syn*-type of addition with the bulky samarium substituent is situated in the thermodynamically formed equatorial position (Scheme 2).

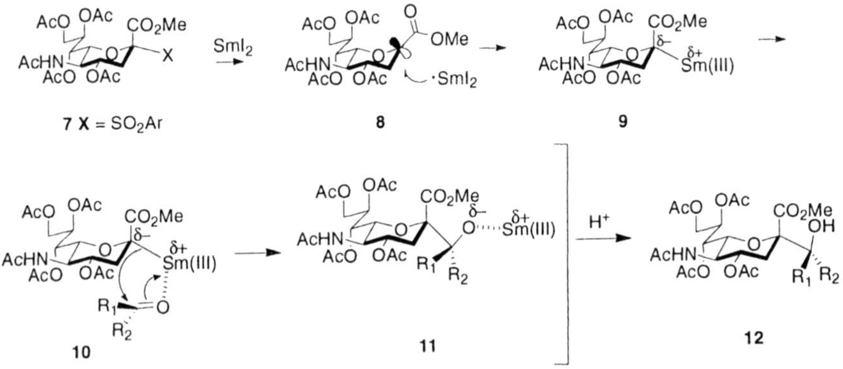

Scheme 2.

When SmI_2 was initially used to mediate the *C*-glycosylation of Neu5Ac pyridyl sulfone **7** donor (*81*) and sugar aldehyde **8** (Scheme 1) hexamethylphosphoric (HMPA) was used. HMPA reportedly accelerates rates of SmI_2 reductions, and it is usually used as a ligand in samarium-mediated reactions (*82*). Neu5Ac α-*C*-glycoside **7** could also be formed in the absence of HMPA (*81,83*), with the characteristic blue color of SmI_2 fading to a cloudy yellow as the reaction proceeded to completion within a few minutes. Furthermore, the Neu5Ac phenyl sulfone donor gave yields identical to the Neu5Ac pyridyl sulfone donor in both the presence and absence of HMPA (Scheme 3). From these studies, it appeared that SmI_2 alone was sufficient for the formation of a relatively stable tertiary intermediate (Scheme 4) at the anomeric center of Neu5Ac. The use of the added HMPA has been conveniently avoided in all further reactions and Neu5Ac phenyl sulfone was used as donor.

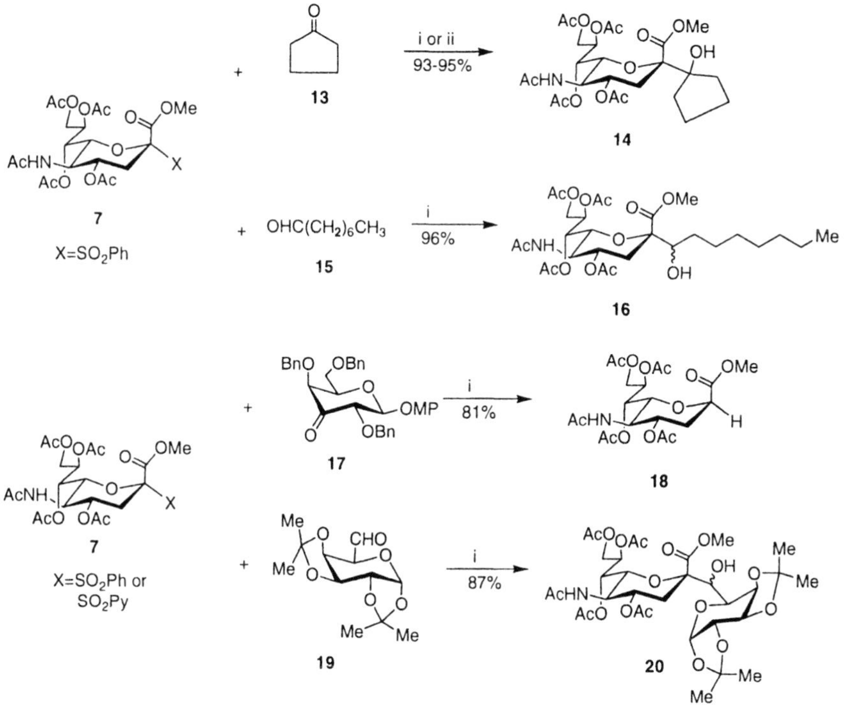

Scheme 3 i. 4 equiv of SmI2 (0.1 M in THF); ii. 4 equiv of SmI2 and 4 equil of HMPA in THF.

When a fully protected sugar ketone **17** (*84*) was subjected to conditions for samarium mediated *C*-glycosylation, no trace of the C-disaccharide was observed. Instead, the α-2-deoxy derivative **18**, corresponding to an inhibitor of X-31 HA hemagglutinin was obtained after deacetylation (*85*). The steric hindrance of sugar ketone **17** might be responsible for the failure of this *C*-glycosylation reaction. The stereochemical outcome of this reaction suggested that the reaction might proceed through an intermediate samarium enolate derivative **21** (Scheme 4). Further mechanistic studies will be required to distinguish between samariated intermediates **10** and **21** (Schemes 2 and 4).

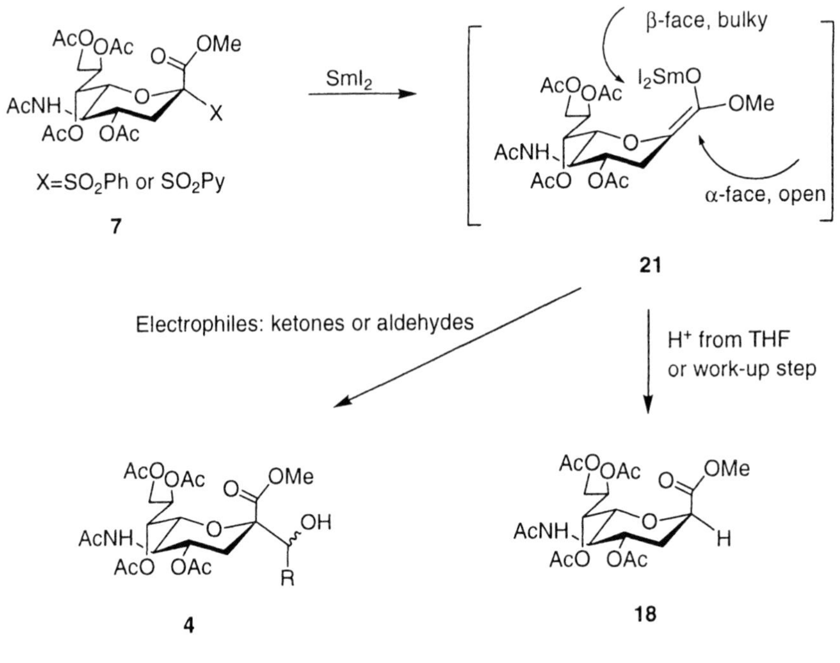

Scheme 4.

Next, the scope of this *C*-glycosylation reaction was investigated. The Neu5Ac chloride was found to serve as a donor in samarium mediated *C*-glycosylation, unfortunately, purification of *C*-glycoside from excess chloride donor is often problematic, making the Neu5Ac phenyl sulfone the donor of choice. A variety of acceptors were also evaluated including alkenes, epoxides, vinyl esters, aldehydes and ketones. Only the aldehydes and ketones afforded the desired *C*-glycoside products.

In conclusion, neuraminic acid 2-pyridyl sulfone, 2-phenyl sulfone or 2-chloro derivatives react with ketones or aldehydes in THF in the presence or absence of HMPA resulting in the near instantaneous and stereospecific formation of Neu5Ac α-*C*-glycosides.

KDN

KDN, 3-deoxy-D-*glycero*-D-*galacto*-2-nonulopyranosylonic acid **2**, is a novel type of sialic acid in which the acetamido group at C-5 of *N*-acetylneuraminic acid is replaced by a hydroxyl group. This ulosonic acid was first isolated from rainbow trout eggs (*86*). In the past 20 years, a number of KDN-glycoconjugates, exhibiting structural determinants related to human tumor-associated antigens, have been reported in mammals (*87-89*). In addition, oligo/poly-KDN and KDN-glycoprotein play an important role in the binding of calcium ions (*90*).

We reported the first stereocontrolled synthesis of KDN containing α-*C*-glycosides *via* glycosyl samarium (III) intermediates (*91-95*). KDN **2** was prepared according to a previously described method (*96*), and the KDN phenylsulfone **22** was synthesized using a similar procedure to that for preparing Neu5Ac phenylsulfone **7** (*97*). Treatment of a neat mixture of KDN phenylsulfone **22** and ketones **13**, **15**, **23**, **25** and **27** (1.2 equiv) in an inert atmosphere with 4 equiv of freshly prepared 0.1 M SmI_2 solution in THF at room temperature, gave a nearly instantaneous conversion to the KDN-*C*-glycosides **14**, **16**, **24**, **26** and **28** in excellent yields (Scheme 5).

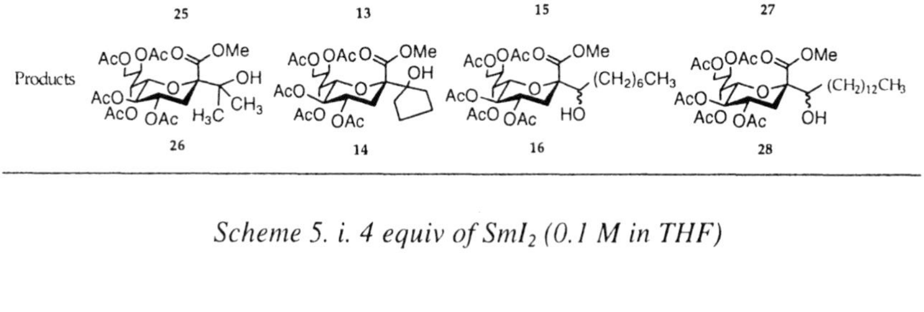

AcO OAc O OMe AcO O X AcO OAc X=SO₂Ph **22** + O= CH₃ CH₃ CH₃ **23** —i→ AcO OAc O OMe OH AcO O AcO OAc **24** CH₃ CH₃ CH₃

Electrophiles: **25** (acetone), **13** (cyclopentanone), **15** OHC(CH₂)₆CH₃, **27** OHC(CH₂)₁₂CH₃

Products: **26**, **14**, **16** ((CH₂)₆CH₃, HO), **28** ((CH₂)₁₂CH₃, OH)

Scheme 5. i. 4 equiv of SmI_2 (0.1 M in THF)

KDN(α2-3)Gal and KDN(α2-6)GalNAc are core structures in the naturally occurring KDN oligosaccharides. Thus, samarium mediated *C*-glycosylation was applied to the synthesis of *C*-linked KDN(α2-3)Gal **29** and KDN(α2-6)Gal **30** derivatives. Coupling of phenylsulfone **22** with sugar electrophile **8** (*98*) under Barbier conditions diastereoselectively generated (2-3) linked *C*-disaccharide **29**, while a similar reaction with **19** (*99*) gave diasteromers at newly formed chiral center (C-6) (Scheme 6).

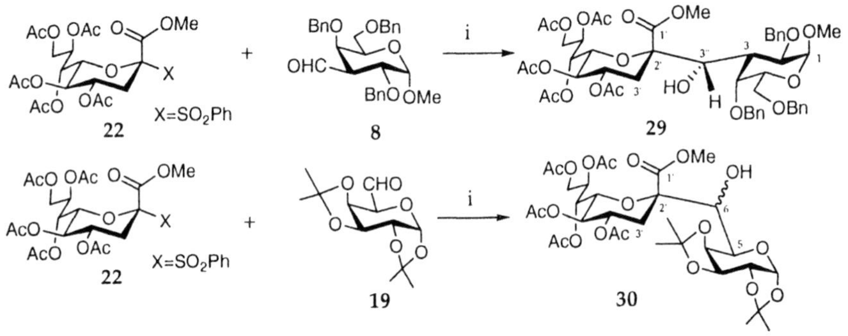

Scheme 6. i. 4 equiv of SmI_2 (0.1 M in THF).

In summary, synthesis of the KDN α-*C* glycosides proceeds from chloride, phenyl sulfone and pyridylsulfone donors, through the same organosamarium intermediate **10** with ketones or aldehydes as observed for the synthesis of Neu5Ac α-*C* glycosides (Scheme 7). This reaction stereospecifically affords tertiary C-C bonds. This chemistry has advantages over previously reported methodologies, employing alkyl lithium species (*100*), since it tolerates acetyl groups and utilizes mild reaction conditions.

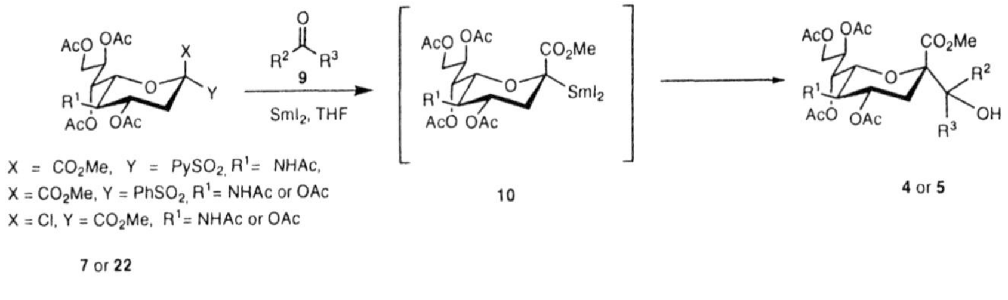

X = CO_2Me, Y = $PySO_2$, R^1= NHAc,
X = CO_2Me, Y = $PhSO_2$, R^1= NHAc or OAc
X = Cl, Y = CO_2Me, R^1= NHAc or OAc

7 or 22

Scheme 7. Synthesis of "C"-glycosides of Neu5Ac and KDN using a ketone or aldehyde containing electrophile. Where R^1 and R^2 are H, alkyl/aryl or saccharide moieties.

KDO

KDO (3-deoxy-D-*manno*-2-octulosonic acid) is a key component of the cell wall lipopolysaccharide (LPS) of Gram-negative bacteria. KDO residues form the critical linkage between the polysaccharide and lipid A regions of LPS (*101-104*). KDO, like Neu5Ac and KDN, is well known to form glycosides in the α-configuration in nature (*105-108*). Our laboratory next set out to use samarium iodide under Barbier conditions to synthesize α-*C*-glycoside of KDO in studies that paralleled those for the successful syntheses of the α-*C*-glycoside Neu5Ac and KDN (Scheme 7). Unfortunately, this chemistry afforded only the undesired β-*C*-glycoside **33** (Scheme 8).

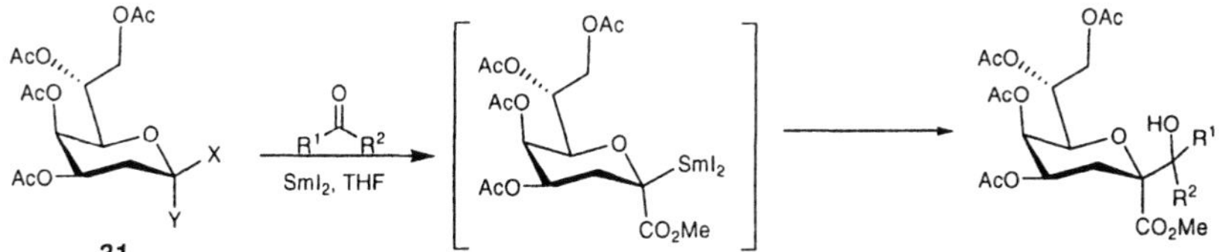

Scheme 8. Scope of samarium chemistry in the synthesis of "C"-glycosides of KDO. Where R^1 and R^2 are H, alkyl or saccharide moieties.

To successfully act as mimics, KDO *C*-glycosides are expected to have the same configuration as the corresponding *O*-glycosides. Unfortunately, unlike Neu5Ac (**1**) and KDN (**2**) which resides primarily in the $^2C_5$ conformation, KDO (**3**) prefers the $^5C_2$ conformation (Figure 1). Thus, a new strategy was necessary to stereoselectively synthesize the α-*C*-disaccharide of KDO. The reaction of *t*-butyl (4,5,7,8-tetra-*O*-acetyl-3-deoxy-α-D-*manno*-2-octulopyranosyl chloride)-onate donor, **36**, with the 6-formylgalactopyranoside acceptor, **19**, in the presence of SmI_2 was expected to form the α-*C*-disaccharide of KDO. The ammonium salt of KDO **34** was prepared according to previously described methods (*105-108*). The synthesis of *t*-butyl (4,5,7,8-tetra-*O*-acetyl-3-deoxy-α-D-*manno*-2-octulopyranosyl chloride)onate **36** was carried out in two steps. The reaction using *t*-butyl trichloroacetimidate (*109-111*) requires a nonpolar solvent, such as cyclohexane, which is unable to dissolve **34**. The acetylation and chlorination of **34** afforded **35**, which was freely soluble in cyclohexane. Esterification of **35** with *t*-butyl trichloroacetimidate in the presence of a catalytic amount of BF_3, gave **36** in 91% yield (2 steps overall). Glycosylation of the 6-formylgalactopyranoside acceptor **19** with the KDO donor **36** in the presence of freshly prepared samarium(II) iodide, afforded the corresponding *C*-disaccharide **37** in 77% yield (Scheme 9a).

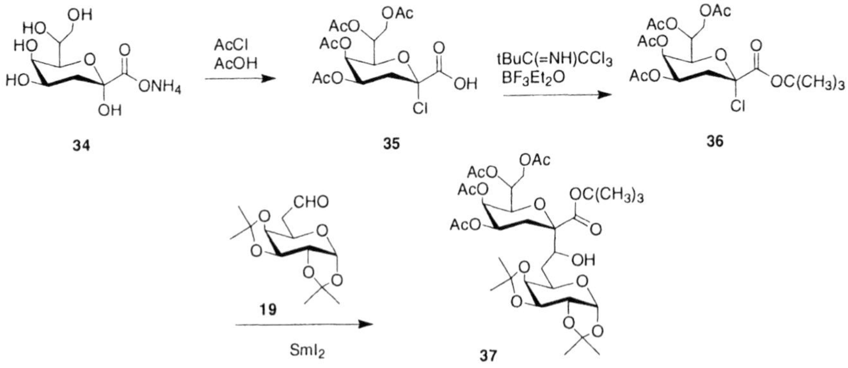

Scheme 9a.

It was rationalized that the KDO *C*-glycoside is generated only in the α-configuration because the α-face of the samarium enolate intermediate **38** is much less sterically hindered than β-face (Scheme 9b).

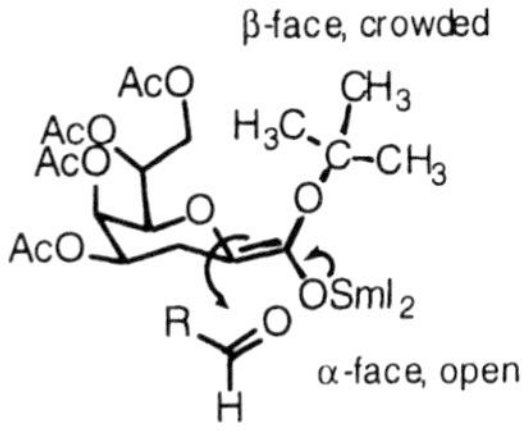

Scheme 9b.

Properties of *O*- and *C*-Glycosides

sTn C-Glycosides

Glycoproteins are major components on the surface of mammalian cells. Many carry *O*-linked oligosaccharides (*O*-glycans), which are conjugated through serine or threonine residues. Others carry *N*-linked oligosaccharides (*N*-glycans), conjugated through an asparagine residue. The recognition of these *O*- and *N*-glycoconjugates play key roles in the transmission of biological information at the cellular level (*112-114*). Their numerous biological functions include roles in cellular recognition, adhesion, cell-growth regulation, cancer cell metastasis, and inflammation. Cell-surface glycans also serve as attachment sites for infectious bacteria, viruses, and toxins, resulting in pathogenesis (*115,116*). Anomalies in cell-surface carbohydrates are often closely associated with cell transformation, malignancy and other various pathological conditions, including immunodeficiency syndromes, cancer and inflammation (*117*).

Sialy Tn (sTn) is a carbohydrate antigen associated with many different types of tumors (*118-120*) as well as viral pathogens, such as HIV (*121*). Danishefsky and coworkers successfully synthesized and evaluated sTn *O*-glycoside and its trimer as cancer vaccines (*112-124*).

Fully protected sTn -*C*- glycoside analog **39** was prepared by C-glycosylation of the neuraminic acid sulfone donor **7** with an aldehyde acceptor **40** (Scheme 10). The donor **7** was prepared from neuraminic acid in four steps as previously reported (*125,126*). The critical intermediate, aldehyde acceptor **40**, was prepared in 14 steps. Starting from the commercially available diisopropylidene galactose derivative **41**, the corresponding 6-iodo derivative **42** was prepared followed by displacement with cyanide nucleophile to yield **43**. The modest yield (30 %) of this reaction might be due to unfavorable electronic and steric effects arising from the ring oxygen atom and the axially oriented oxygen at C-4 respectively. Reduction of the 6-cyano derivative **43** using DIBAL-H afforded aldehyde **44** in moderate overall yield (*127*). Quantitative reduction of aldehyde **44** with $NaBH_4$ in MeOH afforded the corresponding alcohol **45**. De-isopropylidenation of **45** accomplished by treatment with amberlite IR-120 (H^+) resin in water at 80°C for 3 h, provided the 6-deoxy-D-galacto-heptopyranose **46** in quantitative yield. The one carbon extended, galactal **47** was obtained in good yield from **46** using a one pot, three step procedure consisting of: 1. peracetylation with acetic anhydride and catalytic HBr/HOAc; 2. conversion of the anomeric acetate to the corresponding bromide with excess HBr/HOAc; and 3. reductive elimination of the 1-bromo and 2-acetoxy groups with Zn/Cu (*128*). Azidonitration (*129*) of **47** with excess ceric ammonium nitrate (CAN) and sodium azide in dry acetonitrile afforded primarily the 2-azido-1-nitrate addition product having the desired *galacto* configuration. Treatment of the crude product with LiBr (*129*) in dry acetonitrile under ionic conditions afforded **48**.

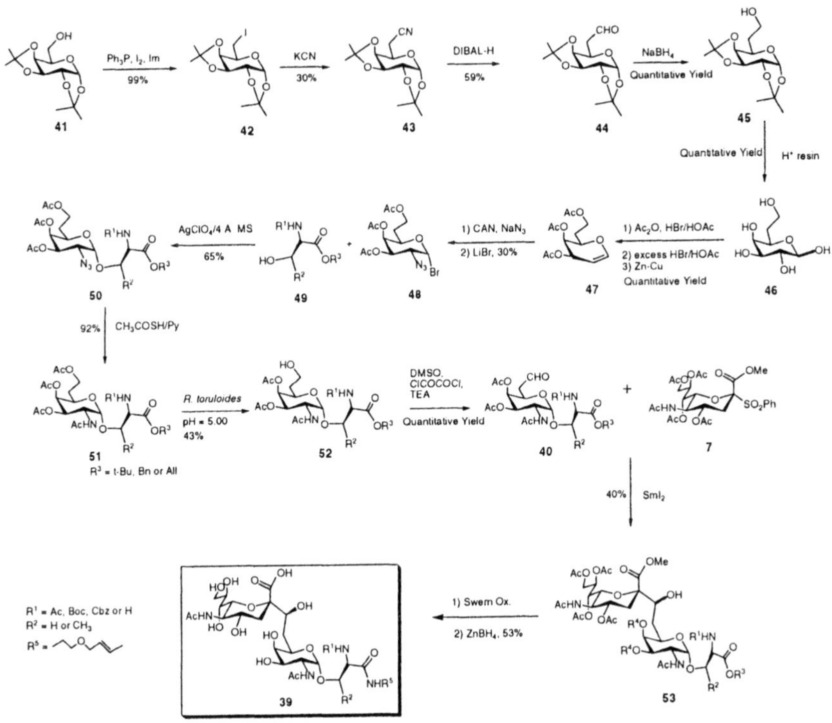

Scheme 10. Synthesis of the C-glycoside analog of sTn.

Glycosylation of **48** with the N^{α} - benzyloxy carbonyl protected OBn ester of L-Serine **49**, prepared as previously reported (*130*), in the presence of silver perchlorate afforded the glycopeptide **50**. Separation of glycopeptide **50** from unreacted starting material, Z-Ser-OBn, was cumbersome since both compounds migrated with similar R_f values on silica. However, conversion of **50** to **51** by treatment with thioacetic acid/pyridine (*131*), resulted in a large change in the glycopeptide polarity, permitting the removal of the unreacted aminoacid acceptor from the desired glycopeptide product **51**. Efforts next focused on the selective deprotection of the C-7 primary acetyl group in **51**. Selective enzymatic deacetylation using an esterase from *Rhodosporidium toruloides* has been previously reported to regioselectively deprotect primary acetates in the presence of secondary acetates (*132-133*). Treatment of **51** with this esterase at pH 5 using a sodium phosphate-sodium citrate buffer afforded **52** in quantitative yield. The site of the enzymatic deacetylation was unequivocally established as C-7. Swern oxidation (*134*) of **52** afforded aldehyde acceptor **40**. Glycosylation of the neuraminic acid sulfone donor **7** with aldehyde acceptor **40** in the presence of freshly prepared SmI_2 (*135-141*) afforded the fully protected sTn α-*C*-glycoside **53** as a diastereomeric mixture. Efforts to deoxygenate the bridge hydroxymethylene group by Barton deoxygenation (*142*) failed. Chemical resolution was achieved by oxidizing the bridge hydroxyl group to ketone which was then stereoselectively reduced (*142*). The pure sTn-α-*C*-glycoside **39** is currently being conjugated to KLH carrier protein for evaluation of efficacy as a carbohydrate vaccine.

Serine-based Neuraminic Acid

The synthesis of serine-based *C*-glycosides has only been carried out in a few laboratories and has relied on cross-metathesis of oxazolidine silyl enol ether (*143-145*); or the Ramberg Bäcklund rearrangement (*146*). Using a method for the preparation of *C*-glycosides of neuraminic acids using SmI_2 (*147*), we have synthesized a serine-based *C*-glycoside of neuraminic acid.

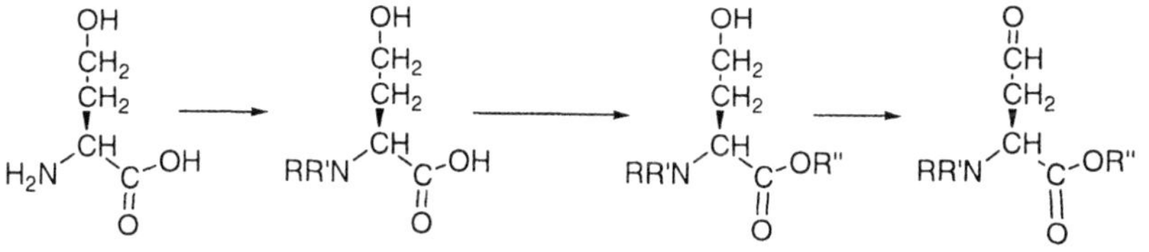

Scheme 11.

To prepare a *C*-glycosidic linkage, it was first necessary to synthesize a neuraminic acid donor **7** (*148*) and a serine-based acceptor **54** (Scheme 11). Homoserine was chosen as the substitute of serine since during *C*-glycosylation, an additional carbon is required to replace the interglycosidic *O*-linkage. The strategy to prepare the acceptor involved the orthogonal protection of amino and carboxyl groups in homoserine and the oxidation of its hydroxyl group to an aldehyde acceptor for *C*-glycosylation. Cbz protected homoserine **56** was converted to corresponding allyl ester **57e** and benzyl ester **57f**, respectively (Scheme 11). Both **57e** and **57f** were smoothly oxidized to aldehyde acceptors **54e** and **54f** (*149*). *C*-glycosylation of these two acceptors afforded *C*-glycoside **58** with concomitant loss of carboxyl protection and subsequent lactonization.

Only a small amount of desired allyl protected homoserine *C*-glycoside **59** was obtained. This intramolecular cyclization product is favorable due to the formation of a stable, five-membered ring, commonly observed in homoserine-based syntheses (Figure 3) (*150*).

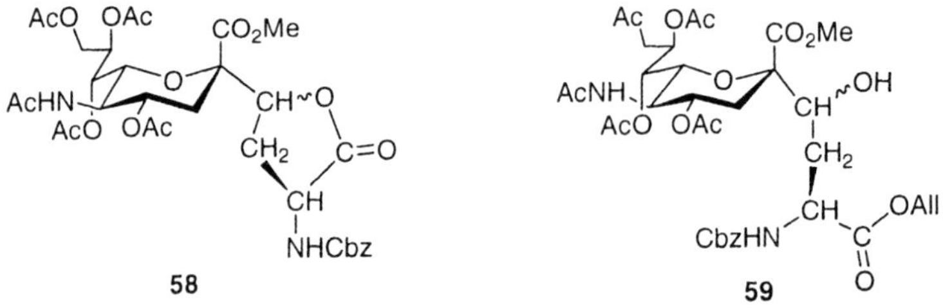

AcO OAc CO2Me AcHN AcO OAc CH~O CH2 C=O CH NHCbz 58

Ac OAc CO2Me AcNH AcO OAc CH OH CH2 CH OAll CbzHN C O 59

Figure 3. Two C-glycosylation products formed.

Lactone **58** could be opened under hydrogenolysis conditions using acetic acid and water as reaction media, affording the unprotected serine-based neuraminic acid *C*-glycoside (*151-152*). Attempts to protect the resulting free amino acid with FmocCl (*153-154*) only afforded the *N*-Fmoc protected, lactonization product.

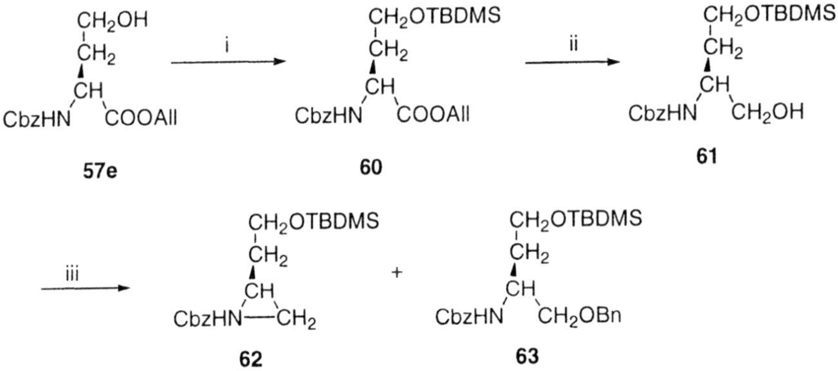

CH_2OH CH_2 CH CbzHN COOAll **57e** →(i) $CH_2OTBDMS$ CH_2 CH CbzHN COOAll **60** →(ii) $CH_2OTBDMS$ CH_2 CH CbzHN CH_2OH **61**

→(iii) $CH_2OTBDMS$ CH_2 CH CbzHN—CH_2 **62** + $CH_2OTBDMS$ CH_2 CH CbzHN CH_2OBn **63**

Scheme 12a. Reagents: (i) TBDMSCl, Pyridine, (ii) LiBH$_4$, THF, (iii) BnBr, 1.2 eq NaH, DMF, 0 ℃

It was clear that the protection of amino group and the presence of an α-COOH group in homoserine complicated samarium-based *C*-glycosylation.

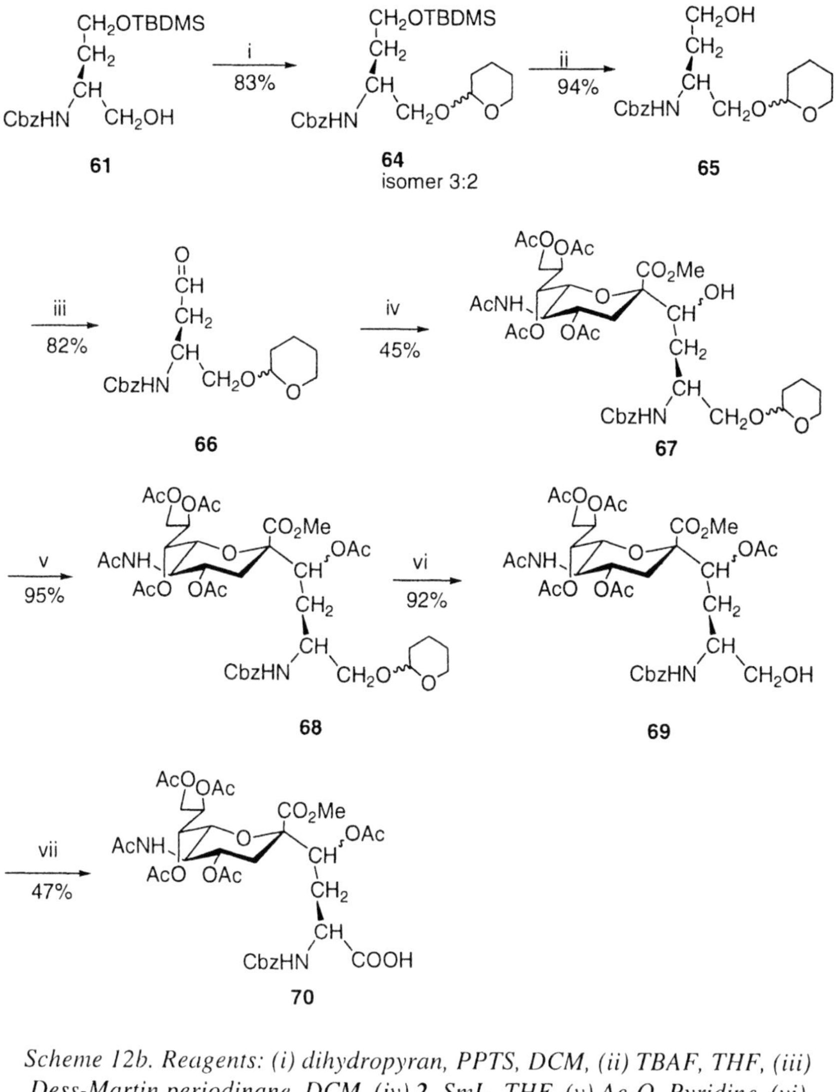

Scheme 12b. Reagents: (i) dihydropyran, PPTS, DCM, (ii) TBAF, THF, (iii) Dess-Martin periodinane, DCM, (iv) **2***, SmI_2, THF, (v) Ac_2O, Pyridine, (vi) PPTS, EtOH, 50 ℃, (vii) $RuCl_3$, $NaIO_4$, CCl_4, CH_3CN, H_2O, 0 ℃.*

Reduction of the carboxyl group at the beginning of the synthesis and its subsequent oxidation following *C*-glycosylation was next examined to avoid lactonization and improve *C*-glycosylation yield.
The hydroxyl group in Cbz-L-homoserine allyl ester **57e** was protected as the silyl ether **60** (155), and the allyl ether was reduced to afford alcohol **61** (Scheme 12a) (*156*). Surprisingly, benzylation of the newly generated hydroxyl group in **61** with benzyl bromide and sodium hydride in DMF at 0 °C afforded aziridine **62** as a major product. Reducing the reaction temperature and the amount of NaH failed to prevent the formation of the undesired aziridine. Benzylation under acidic conditions using silver oxide and benzyl bromide, however, prevented the formation of aziridine **62** but afforded the desired **63** in very low yield (*157*).

Failure of benzyl protection led us to examine tetrahydropyranyl ether (THP) protection, which is easy to install, stable to most nonacidic reagents, and easy to remove. THP protection of **61** went smoothly (*158*) and affording **64** in 83% yield (Scheme 12b). Quantitative tetrabutylamonium fluoride (TBAF) removal of silyl ether afforded **65**, which was readily oxidized to the *C*-glycosylation aldehyde acceptor **66**. *C*-glycosylation of **66** afforded α-*C*-glycoside **67** at 45% isolated yield. Acetylation of newly generated bridge hydroxyl group afforded fully protected *C*-glycoside **68**. Removal of THP using pyridinium *p*-toluensulfonate (PPTS), followed by oxidation of resulting hydroxyl group, generated our target, L-homoserine based *C*-glycoside **70** (*159-161*).

GM4 and GM3

Ganglioside GM4 is an important cell adhesion molecule (*15*). It promotes neuronal adhesion through its interaction with myelin-associated glycoprotein (MAG). This adhesion is important in the regeneration of neuronal tissues (*15*). The C-disaccharide of GM4 **71** is currently being synthesized in our laboratory by standard *O*-glycosylation of C-disaccharide **72** (Scheme 13) with protected ceramic acid **74** prepared by the method previously described by Hasegawa (*162*). The *C*-disaccharide donor **72** has been successfully synthesized (Scheme 14) from perbenzylated 3-formyl galactose acceptor **73** and Neu5Ac phenyl sulfone donor **7**. Once the ceramide has been attached the ganglioside analog will be deprotected to afford GM4 *C*-glycoside **71**. The ceramide side chain of the *C*-glycosides of GM4 **71** will be oxidized with ozone to afford an aldehyde group that was coupled to KLH by reductive amination (*163*). This conjugate will be used by our laboratory to prepare antibodies in mice against *C*-GM4 that was tested for cross-reactivity with the natural GM4 (*O*-glycoside). The solution conformation of *C*-GM4 will also be compared to GM4 using NMR spectroscopy (*111*).

GM3, a related trisaccharide containing ganglioside, plays an important role in cell growth and differentiation. GM3 inhibits epidermal growth factor (EGF) receptor mediated signal transduction (*164*). Tumors, such as those involved in brain cancer, overexpress EGF receptor. Effective new therapeutic agents might modulate the mitogenic effect of EGF on such tumors. The synthesis of *C*-oligosaccharide GM3 **93** is currently underway starting from lactose **94** (Scheme 15). The planned chemical synthesis of *C*-GM3 **93** follows a parallel route to that currently being successfully used in *C*-GM4 synthesis. Once synthesized, the conformational flexibility and biological activity of *C*-GM3 **93** will be evaluated.

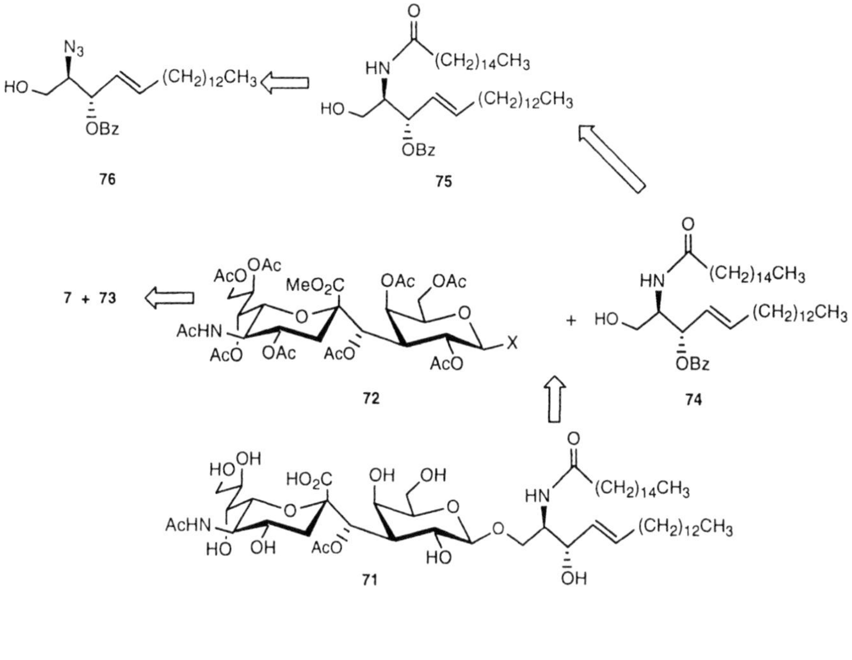

Scheme 13. Retrosynthesis of "C"-glycoside of GM4 **71**.

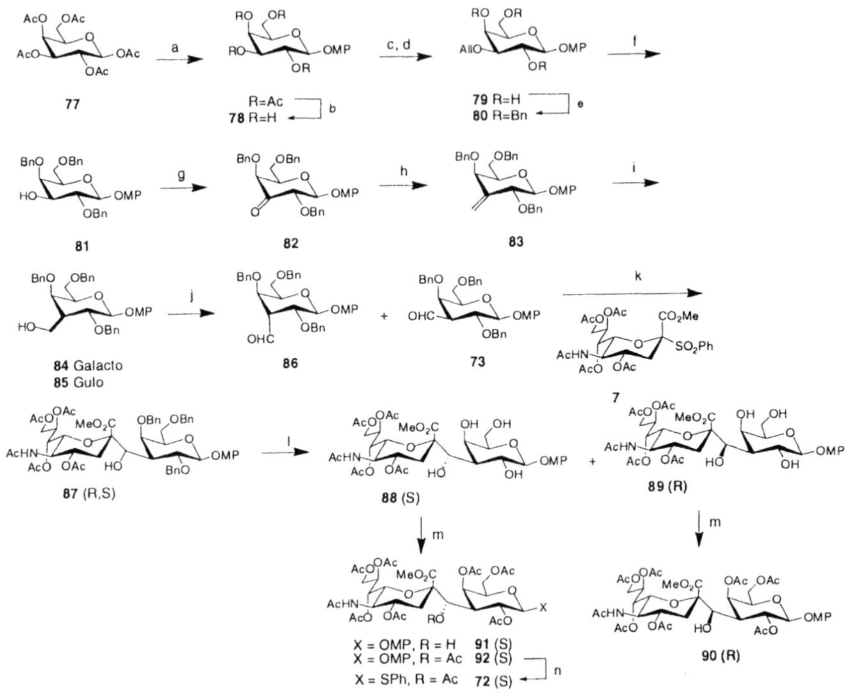

Scheme 14. Synthesis of the GM4 C-disaccharide **72**. Reagents: (a) TMSTf, MeOPhOH, CH_2Cl_2, (b) NaOMe, MeOH, (c) Bu_2SnO, MeOH, DMF, reflux, (d) AllBr, Bu_4NI, toluene, (e) BnBr, NaH, DMF, (f) $PdCl_2$, MeOH, toluene, (g) DMSO, Ac_2O, (h) Tebbe's reagent, THF, (i) 9-BBn, THF, H_2O_2, NaOH, (j) DMSO, $(COCl)_2$, CH_2Cl_2, (k) SmI_2, THF, (l) H_2, Pd/C, EtOAc, CH_3OH, H_2O, AcOH, (m) Ac_2O, Pyridine, CH_3OH, (n) , C_6H_5SH, BF_3OEt_2, toluene.

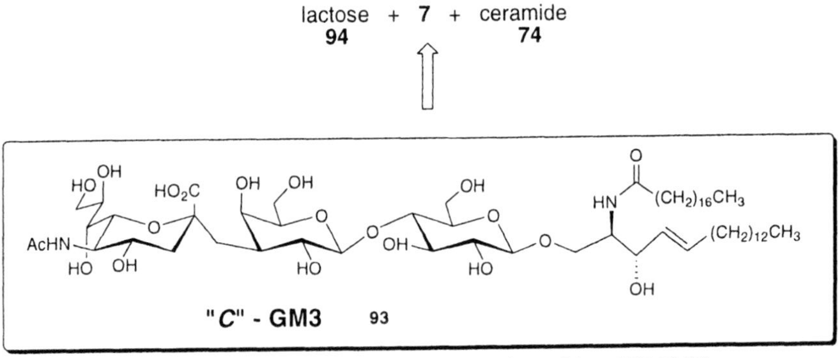

Scheme 15. Retrosynthesis of "C"-glycoside of GM3 **93**.

Conclusions

In conclusion, samarium mediated *C*-glycosylation is an effective method of synthesizing α-*C*-glycosides of Neu5Ac, KDN, and KDO. These *C*-glycosides can be incorporated into structures corresponding to natural products of biological interest affording catabolically stable analogues of a number of potentially important biological activities. For example, recent evaluation of Neu5Ac with hydrophobic aglycones demonstrate reasonably favorable activity as neuraminidase inhibitors (*165-166*). Moreover, conformational evaluation of Neu5Ac *C*-glycosides show that they can occupy similar conformational space as the *O*-glycoside natural products (*167*). Thus, these *C*-glycosides should both resemble the natural products and provide analogs with excellent stability. Only future biological testing will provide the results needed to confirm the value of analogs.

Experimental

General procedure for Neu5Ac, KDN or KDO *C*-glycosidation.

An appropriately hydroxyl (*i.e.*, acetylated) and carboxyl (*i.e.*, methyl esterified) protected Neu5Ac, KDN or KDO phenyl sulfone glycoside (2 to150 mg) and 1.2 to 2.0 equiv of electrophile (*i.e.*, a ketone or aldehyde containing molecule) are dried together under high vacuum for 4 h, then dissolved in degassed anhydrous THF (0.5 to1 mL). SmI_2 (4 equiv, freshly prepared from Sm and ICH_2CH_2I, 0.1M in THF) is added in one portion at room temperature with vigorous stir-ring. After approximately10 min, the reaction mixture is directly filtered, and the filtrate is concentrated under reduced pressure and then purified on silica gel column with the appropriate eluent such as EtOAc.

References

1. Schauer, R. *Adv. Carbohydr. Chem. Biochem.* **1982,** *40,* 131-234.
2. *Sialic Acids, Chemistry, Metabolism and Function. Springer Verlag;* Schauer, R., Ed.; Wien: New York, NY, 1982.
3. Jeanloz, R.; Codingten, J. *The biological role of sialic acid at the surface of the cell. In: Biological Roles of Sialic Acids;* Rosenberg, A.; Schengrund, C.; Eds.; Plenum Press: New York, NY, 1976, 201-238.
4. Varki, A. *Glycobiology* **1992,** 2, 25-40.

5. Kleuk, E.; Gottschalk, A. *Chemistry and Biology of Mucopolysaccharides;* Wolstenholme, G.; O'Connor, M.; Eds.; Little Brown: Boston, MA, 1958, 306.
6. Kemp, R.B. *J. Cell. Sci.* **1970,** 6, 751-766.
7. Ahmad, F.; McPhie, P. *Int. J. Biochem.* **1980,** 11, 91-96.
8. Rosen, S.R. *Immunology* **1993,** 5, 237-247.
9. McEver, R.P. *Current Opinion in Immunology* **1994,** 6, 75-84.
10. Nelson, R.M.; Venot, A.; Bevilacqua, M.P.; Linhardt, R.J.; Stamenkovic, I. *Ann. Rev. Cell Dev. Biol.* **1995,** 11, 601-631.
11. Paulson, J.; Rogers, G.; Carroll, M.; Higa, S.; Pritchett, T.; Milks, G.; Sabesan, S. *Pure Appl. Chem.* **1984,** 56, 797-805.
12. Choppin, P.; Scheid, A. *Rev. Infect. Dis.* **1980,** 2, 40-61.
13. Schauer, R. *TIBS* **1985,** 10, 357-363.
14. Schauer, R. *Pure Appl. Chem.* **1984,** 56, 907-921.
15. Yang, L.J.; Zeller, C.B.; Shaper, N.L.; Kiso, M.; Hasegawa, A.; Shapiro, R.E.; Schnaar, R.L. *Proc. Nat. Acad. Sci. (USA)* **1996,** 93, 814-818.
16. Swartley, J.S.; Marfin, A.A.; Edupugantis, D.; Liu, L.J.; Cieslak, P.; Perkins, B.; Wenger, J.D.; Stephans, D.S. *Proc. Nat. Acad. Sci. (USA)* **1997,** 94, 271-276.
17. Colman P, Neuraminidase: enzyme and antigen. In: The Influenza Viruses, Krug R (Ed.), Plenum Press, New York, 1989, 175-218.
18. Jennings, H.J. *Adv. Carbohydr. Chem. Biochem.* **1983,** 41, 155-208.
19. Kenne, L.; Lindberg, B. *Bacterial polysaccharides. In The Polysaccharides;* Aspinall, G.O.; Ed.; Academic Press: New York, NY, 1983, Vol 2, 287-363.
20. Geyer, H.; Holschback, C.; Hunsmann, G.; Schneider, J. *J. Biol. Chem.* **1988,** 263, 11760-11767.
21. Inoue, S.; Kitajima, K.; Inoue, Y. *J. Biol. Chem.* **1996,** 271, 24341-24344.
22. Edebrink, P.; Jansson, P.E.; Bogwald, J.; Hoffman, J. *Carbohydr. Res.* **1996,** 287, 225-245.
23. Stepanova, L.K.; Belaia, IuA.; Sergeeva, N.S.; Ageeva, V.A.; Petrukhin, V.G. *Zh. Mikrobiol. Epidemiol.* **1993,** 1, 51-56.
24. Thomas, J.M.; Camprubi, S.; Merino, S.; Davey, M.R.; Williams, P. *Infect. Immun.* **1991,** 59, 2006-2011.
25. Kaijser, B.; Jodal, U. *J. Clin. Microbiol.* **1984,** 19, 264-266.
26. Trimble, R.B.; Tarentino, A.L. *J. Biol. Chem.* **1991,** 266,1646-1651.
27. Arentino, A.L.; Gomez, C.M.; Plummer, T.H. Jr. *Biochem.* **1985,** 24, 4665-71.
28. Elder, J.H.; Alexander. S. *Proc. Nat. Acad. Sci. (USA).* **1982,** 79, 4540-45404.
29. Ghosh, P.; Bachhawat, B.K.; Surolia, A. *Arch. Biochem. Biophys.* **1981,** 206, 454-457.

30. Baenziger, J.U.; Maynard, Y. *J. Biol. Chem.* **1980,** 255, 4607-4613.
31. Hassler, D.F.; Bell, R.M. *Adv. Lipid Res.* **1993,** 26, 49-57.
32. Haeuw, J.F.; Michalski, J.C.; Strecker, G.; Spik, G.; Montreuil, J. *Glycobiology* **1991,** 1, 487-492.
33. Baussant, T.; Strecker, G.; Wieruszeski, J.M.; Montreuil, J.; Michalski, J.C. *Eur. J. Biochem.* **1986,** 159, 381-385.
34. Montreuil, J. *Comptes. Rendus.* **1981,** 175, 694-707.
35. Eisele, T.; Toepfer, A.; Kretzschmar, G.; Schmidt, R. R. *Tetrahedron Lett.* **1996,** 37, 1389-1392.
36. Hasegawa, A.; Terada, T.; Ogawa, H.; Kiso, M. *J. Carbohydr. Chem.* **1992,** 11, 319-331.
37. Ress, D. K.; Linhardt, R. J. *Curr. Org. Synthesis* **2004,** 1, 31-46.
38. Hasegawa, A.; Ohki, H.; Nagahama, T.; Ishida, H.; Kiso, M. *Carbohydr. Res.* **1991,** 212, 277-281.
39. Kitajima, K.; Kuroyanagi, H.; Inoue, S.; Ye, J.; Troy, F.A.; Inoue, Y. *J. Biol. Chem.* **1994,** 269, 21415-21419.
40. Du, Y.; Linhardt, R.J.; Vlahov, I.R. *Tetrahedron* **1998,** 54, 9913-9959.
41. Kanie, O.; Kiso, M.; Hasegawa, A. *J. Carbohydr. Chem.* **1988,** 7, 501-506.
42. Levy, D.E.; Tang, C. *The chemistry of C-glycosides;* Pergamon, Elsevier Science, Ltd. 1995.
43. Hung, S.C., Wong, C.-H. *Angew. Chem. Int. Ed. Engl.* **1996,** 35, 2671-2674.
44. Sinaÿ, P. *Pure & Appl. Chem.* **1997,** 69, 459-463.
45. Beau, J.-M.; Gallagher, T. *Topics Curr. Chem.* **1997,** 187, 1-54.
46. Paulsen, H.; Matschulat, P. *Liebigs. Ann. Chem.* **1991,** 487-495.
47. Walliman, K.; Vasella, A. *Helv. Chim. Acta.* **1991,** 74, 1520-1532.
48. Nagy, J.; Bednarski, M. *Tetrahedron Lett.* **1991,** 32, 3953-3956.
49. Luthman, K.; Orbe, M.; Waglund, T.; Claesson, A. *J. Org. Chem.* **1987,** 52, 3777-3784.
50. Suzuki, J.; Murakami, K.; Nishimura, Y. *J. Carbohydr. Chem.* **1993,** 12, 201-208.
51. Jedrzejas, M.J.; Singh, S.; Brouillette, W.J.; Air, G.M.; Luo, M. *Proteins* **1995,** 23, 264-277.
52. Szkudlinski, M.W.; Thotakura, N.R.; Weintraub, B.D. *Proc. Acad. Sci. (USA)* **1995,** 92, 9062-9066.
53. Lee, Y.C. *Biochem. Soc. Trans.* **1993,** 21, 460-463.
54. Rice, K.G.; Lee, Y.C. *Adv. Enzymol. Relat. Areas Molec. Biol.* **1993,** 66, 41-83.
55. Rice, K.G.; Chiu, M.H.; Wadhwa, M.S.; Thomas, V.H.; Stubbs, H.J. *Adv. Exp. Med. Biol.* **1995,** 376, 271-282.
56. Livingston, P.O. *Curr. Opin. Immunol.* **1992,** 4, 624-629.
57. MacLean, G.D.; Longenecker, B.M. *Can. J. Oncol.* **1994,** 4, 249-254.

58. Espinosa, J.-F.; Cañada, F.J.; Asensio, J.L.; Dietrich, H.; Martín-Lomas, M.; Schmidt, R.R.; Jiménez-Barbero, J. *Angew. Chem. Int. Ed. Engl.* **1996,** 35, 303-306.
59. Mouritsen, S.; Meldal, M.; Christiansen-Brams, I.; Elsner, H.; Werdelin, O. *Eur. J. Immunol.* **1994,** 24, 1066-1072.
60. Livingston, P.O. *Cancer Biology* **1995,** 6, 357-366.
61. Deng, S.-J.; Mackenzie, C.R.; Hirama, T.; Brousseau, R.; Lowary, T.L.; Yound, N.M.; Bundle, D.R.; Narang, S.A. *Proc. Nat. Acad. Sci. (USA)* **1995,** 92, 4992-4996.
62. Ragupathi, G.; Koganty, R.R.; Qiu, D.; Lloyd, K.O.; Livingston, P.O. *Glycoconjugate J.* **1998,** 15, 217-221.
63. Livingston, P.O. *Ann. New. York. Acad. Sci.* **1993,** 690, 204-213.
64. Wei, A.; Kishi, Y. *J. Org. Chem.* **1994,** 59, 88-96.
65. Xu, Q.; Gitti, R.; Bush, C.A. *Glycobiology*, **1996,** 6, 281-288.
66. Espinosa, J.-F.; Cañada, F.J.; Asensio, J.L.; Martín-Pastor, M.; Dietrich, H.; Martín-Lomas, M.; Schmidt, R.R.; Jiménez-Barbero, J. *J. Am. Chem. Soc.* **1996,** 118, 10862-10871.
67. Espinosa, J.-F.; Martín-Pastor, M.; Asensio, J.L.; Dietrich, H.; Martín-Lomas, M.; Schmidt, R.R.; Jiménez-Barbero, J. *Tetrahedron Lett.* **1995,** 36, 6329-6332.
68. Poveda, A.; Asensio, J.L.; Martín-Pastor, M.; Jiménez-Barbero, J. *Chem. Commun.* **1996,** 3, 421-422.
69. Coterón, J.M.; Singh, K.; Asensio, J.L.; Domínguez-Dalda, M.; Fernández-Mayoralas, A.; Jiménez-Barbero, J.; Martín-Lomas, M. *J. Org. Chem.* **1995,** 60, 1502-1519.
70. Pellequer, J.L.; Westhof, E. *J. Molec. Graphics* **1993,** 11, 204-210.
71. Nores, G.A.; Dahi, T.; Taniguchi, M.; Hakomori, S. *J. Immunol.* **1987,** 139, 3171-3176.
72. Vlasova, E.V.; Vorozhaikina, M.M.; Khraltsova, L.S.; Tuzikov, A.B.; Popova, I.S.; Tsvetkov, IuE.; Nifantev, N.E.; Bovin, N.V. *Bioorganischeskaia Khimiia.* **1996,** 22, 358-365.
73. Winn, R.K.; Sharar, S.R.; Vedder, N.B.; Harlan, J.M. *Ciba Found Symp.* **1995,** 189, 63-71.
74. Sauter, N.K.; Glick, G.D.; Crowther, R.L.; Park, S.J.; Eisen, M.B.; Skehel, J.J.; Knowles, R.J.; Wiley, D.C. *Proc. Nat. Acad. Sci. (USA)* **1992,** 89, 324-328.
75. Toogood, P.L.; Galliker, P.K.; Glick, G.D.; Knowles, J.R. *J. Med. Chem.* **1991,** 34, 3138-3140.
76. Harms, G.; Reuter, G.; Corfield, A.P.; Schauer, R. *Glycoconj. J.* **1996,** 13, 621-630.
77. Herrler, G.; Gross, H.J.; Brossmer, R. *Biochem. Biophys. Res. Commun.* **1995,** 216, 821-827.

78. Michalides, R.; Kwa, B.; Springall, D.; VanZandwijk, N.; Koopman, J.; Hilknes, J.; Mooi, W. *Int. J. Cancer* **1994,** 8, 34-37.
79. Pouilly, P.; Chénedé, A.; Mallat, J.-M.; Sinaÿ, P. *Bull. Soc. Chim. Fr.* **1993,** 130, 256-265.
80. Hung, S.-C.; Wong, C.-H. *Angew. Chem. Int. Ed. Engl.* **1996,** 35, 2671-2674.
81. Vlahov, I.R.; Vlahova, P.I.; Linhardt, R.J. *J. Am. Chem. Soc.* **1997,** 119, 1480-1481.
82. Molander, G.A.; Harris, C.R. *Chem. Rev.* **1996,** 96, 307-338.
83. Marra, A.; Sinaÿ, P. *Carbohydr. Res.* **1989,** 187, 35-42.
84. Schmidt, R. R.; Beyerbach, A. *Liebigs Ann. Chem.* **1992,** 983-986.
85. Bandgar, B.P.; Hartmann, M.; Schmid, W.; Zbiral, E. *Liebig Ann. Chem.* **1990,** 1185-1195.
86. Nadano, D.; Jwasaki, M.; Endo, S.; Kitajima, K.; Inoue, S.; Inoue, Y. *J. Biol. Chem.* **1986,** 261, 11550-11557.
87. Plancke, Y.; Wieruszeski, J.-M.; Alonso, C.; Boilly, B.; and Strecker, G. *Eur. J. Biochem.* **1995,** 231, 434-439.
88. Maes, E.; Plancke, Y.; Delplace, F.; Strecker, G. *Eur. J. Biochem.* **1995,** 230, 146-156.
89. Moolenaar, C. E.; Muller, E. J.; Schol, D. J.; Figdor, C. G.; Bock, E.; Bitter-Suermann, D.; Michalidas, R. J. *Cancer Res.* **1990,** 50, 1102-1106.
90. Shimoda, Y.; Kitajima, K.; Inoue, S.; Inoue, Y. *Biochemistry* **1994,** 33, 1202-1208.
91. Vlahov, I. R.; Vlahova, P. I.; Linhardt, R. J. *J. Am. Chem. Soc.* **1997,** 119, 1480-1481.
92. Mazeas, D.; Skrydstrup, T.; Beau, J.-M. *Angew. Chem. Int. Ed. Engl.* **1995,** 34, 909-912.
93. Namy, J.; Collin J., Bied, C.; Kagan, H. *Synlett.* **1992,** 9, 733-734.
94. Curran, D.; Fevig, T.; Jasperse, C.; Totleben, M.; *ibid*, 943-961, 1992.
95. Molander, G.; McKie, J. *J. Org. Chem.* **1991,** 56, 4112-4120.
96. Nakamura, M.; Furuhata, F.; Yamasaki, T.; Ogura, H. *Chem. Pharm. Bull.* **1991,** 39, 3140-3144.
97. Marra, A.; Sinaÿ, P. *Carbohydr. Res.* **1989,** 187, 35-42.
98. Schmidt, R. and Beyerbach, A. *Liebigs Ann. Chem.* **1992,** 983-986.
99. Dess, D. B.; Martin, J. C. *J. Am. Chem. Soc.* **1991,** 113, 7277-7287.
100. Crich, D.; Lim, L. B. L. *J. Chem. Soc. Perkin Trans. I.* **1991,** 2209-2214.
101. Dumanski, A. J.; Hedelin, H.; Edinliljegren, A.; Beauchemin, D.; Mclean, R. J. C. *Infect. Immun.* **1994**, 62, 2998.
102. Velasco, J.; Moll, H.; Knirel, Y. A.; Sinnwell, V.; Moriyón, I.; Zähringer, U. *Carbohydr. Res.* **1998**, 306, 283-290.
103. Müller-Loennies, S.; Holst, O.; Lindner, B.; Brade, H. *Eur. J. Biochem.* **1999**, 260, 235-249.

104. Vinogradov, E.; Sidorczyk, Z. *Carbohydr. Res.* **2000**, 326, 185-193.
105. Unger, F. M. *Adv. Carbohydr. Chem. Biochem.* **1981**, 38, 323.
106. Schauer, R. *Adv. Carbohydr. Chem. Biochem.* **1982**, 40, 131.
107. Reglero, A.; Rodriguezaparicio, L. B.; Luengo, J. M. *Int. J. Biochem.* **1993**, 25, 1517.
108. Schauer, R.; Kelm, S.; Roggentin, G.; Shaw, L. *Biochemistry and Role of Sialic acids;* Plenum Press: New York, NY, 1995.
109. Vlahov, I. R.; Vlahov, P. I.; Linhardt, R. J. *J. Am. Chem. Soc.* **1997**, 119, 1480-1481.
110. Bazin, H. G.; Du, Y.; Polat, T.; Linhardt, R. J. *J. Org. Chem.* **1999**, 64, 7254-7259.
111. Poveda, A.; Asensio, J. L.; Polat, T.; Bazin, H. G.; Linhardt, R. J.; Jiménez-Barbero, J. *Eur. J. Org. Chem.* **2000**, 1805-1813.
112. Saxon, E.; Bertozzi, C. R. *Annu. Rev. Cell. Dev. Biol.* **2001**, 17, 1-23.
113. Hughes, R. C. *Glycoconjugate J.* **2001**, 17, 567.
114. Bertozzi, C. R.; Kiessling, L. L. *Science* **2001**, 291, 2357-2364.
115. Dwek, R. A. *Chem. Rew.* **1996**, 96, 683-720.
116. Varki, A. *Glycobiology* **1993**, 3, 97.
117. Tsuboi, S.; Fukuda, M. *Bioessays* **2001**, 23, 46.
118. Schauer, R. *Trends Glycosci. Glyc.* **1997,** 9, 315-330.
119. Traving, C.; Schauer, R. *Cell. Mol. Life Sci.* **1998,** 54, 1330-1349.
120. Varki, A. *Glycobiology* **1992,** 2, 25-40.
121. Air, G. M. ; Laver, W. *Proteins: Struct. Funct. Genet.* **1989,** 6, 341-356.
122. Vlahov, I.R.; Vlahova, P.I.; Linhardt, R.J. *J. Am. Chem. Soc.* **1997,**119, 1480-148.
123. Bazin, H.G.; Du, Y.; Polat, T.; Linhardt, R.J. *J. Org. Chem.* **1999,** 64, 7254-7259.
124. Du, Y.; Polat, T.; Linhardt, R.J. *Tetrahedron Lett.* **1998,** 39, 5007-5010.
125. Cao, S.; Meuneir, S.; Andersson, F.; Letellier, M.; Roy, R. *Tetrahedron Asymm.* **1994,** 5, 2303-2312.
126. Marra, A.; Sinaÿ, P. *Carbohydr. Res.* **1989,** 187, 35-42.
127. Du, Y. ; Linhardt, R. J. *J. Carbohydr. Res.* **1998,** 308, 161-164.
128. Shull, B.K.; Wu, Z.; Koreeda, M. *J. Carbohydr. Chem.* **1996,** 15, 955-964.
129. Lemieux, R.U.; Ratcliffe, R.M. *Can. J. Chem.* **1979,** 57, 1244-1251.
130. Schultz, M.; Kunz, H. *Tetrahedron-Asymm.* **1993,** 4, 1205-1220.
131. Rosen, T.; Lico, I. M.; Chu, D. T. W. *J. Org. Chem.* **1988,** 53, 1580-1582.
132. Horrobin, T.; Tran, C.H.; Crout, D. *Perkin Trans.* **1998,** 1, 1069-1080.
133. Kuberan, B.; Wang, Q.; Koketsu, M.; Linhardt, R.J. *Syn. Commun.* **2002,** 32, 1421-1426.
134. Tidwell, T.T. *Synthesis* 1990, 857-870.
135. Namy, J.L.; Collin, J.; Bied, C.; Kagan, H.B. *Synlett.* **1992,** 733-734.
136. Molander, G.; McKie, J. *J. Org. Chem.* **1991,** 56, 4112-4120.

137. Miquel, N. ; Doisneau, G. ; Beau, J.-M. *Angew. Chem. Int. Ed. Engl.* **2000,** 39, 4111-4114.
138. Du, Y. ; Linhardt, R. J. *Carbohydr. Res.* **1998,** 308, 161-164.
139. Molander, G.A.; Harris, C.R. *Chem. Rev.* **1996,** 96, 307-338.
140. Molander, G.A.; Harris, C.R. *Tetrahedron* **1998,** 54, 3321-3354.
141. Krief, A.; Laval, A-M. *Chem. Rev.* **1999,** 99, 745-777.
142. Kuberan, B.; Sikkandar, S.A.; Tomiyama H.; Linhardt, R.J. *Angewandte Chemie* in press, 2003.
143. Dondoni, A.; Marra, A. *Chem. Commun.* **1998**, 16, 1741-1742.
144. Dondoni, A.; Giovannini, P. P.; Marra, A. *J. Chem. Soc. Perkin. Trans. 1* **2001**, *19*, 2380.
145. Dondoni, A.; Mariotti, G.; Marra, A.; Massi, A. *Synthesis* **2001**, 14, 2129-2137.
146. Ohnishi, Y.; Ichikawa, Y. *Bioorg. Med. Chem. Lett.* **2002**, 12, 997-999.
147. Vlahov, I. R.; Vlahov, P. I.; Linhardt, R. J. *J. Am. Chem. Soc.* **1997**, 119, 1480-1481.
148. Marra, A.; Sinaÿ, P. *Carbohydr. Res.* **1989**, 187, 35-42.
149. Ricci, M.; Madariaga, L.; Skrydstrup, T. *Angew. Chem. Inter. Ed.* **2000**, 39, 242-246.
150. Dess, D. B.; Martin, J. C. *J. Am. Chem. Soc.* **1991**, 113, 7277-7287.
151. Ozinskas, A. J.; Rosenthal, G. A. *J. Org. Chem.* **1986**, 51, 5047-5050.
152. Nardo, C. D.; Varela, O. *J. Org. Chem.* **1999**, 64, 6119-6125.
153. Fleet, G. W. J.; Ramsden, N. G.; Witty, D. R. *Tetrahedron* **1989**, 45, 327-336.
154. Hilaire, P. M. S.; Cipolla, L.; Franco, A.; Tedebark, U.; Tilly, D. A.; Meldal, M. *J. Chem. Soc., Perkin Trans. 1* **1999**, 3559.
155. Carpino, L. A.; Han, G. Y. *J. Org. Chem.* **1972**, 37, 3404-3409.
156. Jonghe, S. D.; Lamote, I.; Venkataraman, K.; Boldin, S., A.; Hillaert, U.; Rozenski, J.; Hendrix, C.; Busson, R.; Keukeleire, D. D.; Calenbergh, S. V.; Futerman, A. H.; Herdewijn, P. *J. Org. Chem.* **2002**, 67, 988-996.
157. Laib, T.; Chastanet, J.; Zhu, J. *J. Org. Chem.* **1998**, 63, 1709-1713.
158. Saotome, C.; Kanie, Y.; Kanie, O.; Wong, C.-H. *Bioorg. Med. Chem.* **2000**, 8, 2249-2261.
159. Miyashita, N.; Yoshikoshi, A.; Grieco, P. A. *J. Org. Chem.* **1977**, 42, 3772-3774.
160. Trost, B. M.; Krueger, A. C.; Bunt, R. C.; Zambrano, J. *J. Am. Chem. Soc.* **1996**, 118, 6520-6521.
161. Sham, H. L.; Stein, H.; Cohen, J. *J. Chem. Soc., Chem. Commun.* **1987**, 23, 1792.
162. Kiso, M.; Nakamura, A.; Tomita, Y.; Hasegawa, A. *Carbohydr. Res.* **1986,** 158, 101-111.

163. Helling, F.; Shang, A.; Calves, M.; Zhang, S.; Ren, S.; Yu, R.K.; Oettgen, H.F.; Livingston, P.O. *Cancer Res.* **1994,** 54,197-203.
164. Rebbaa, A.; Hurh, J.; Yamamoto, H.; Kersey, D.S.; Bremer, E.G. *Glycobiology* **1996,** 6, 399-406.
165. Wang, Q.; Wolff, M.; Polat, T.; Linhardt, R.J. *Bioorg. Med. Chem. Lett.* **2000,** 10, 941-944.
166. Wang, Q.; Dordick, J.S.; Linhardt, R.J. *Org. Lett.* **2003,** 5, 1187-1189.
167. Poveda, A.; Asensio, J.L.; Polat, T.; Bazin, H.; Linhardt, R.J.; Jiménez-Barbero, J. *Eur. J. Org. Chem.* **2000,** 1805-1813.

Chapter 4

Sythesis of *C*-Glycosides with Glycosyl Phosphates

Emma R. Palmacci[1,2], Holger Herzner[1], and Peter H. Seeberger[1,3]

[1]**Department of Chemistry, Massachusetts Institute of Technology, 77 Massachusetts Avenue, Cambridge, MA 02139**
[2]**Thios Pharmaceuticals, 5980 Horton, Suite 400, Emeryville, CA 94608**
[3]**Laboratorium fuer Organische Chemie, HCI F 315 Wolfgang-Pauli-Strasse, 10 ETH-Hőnggerberg, CH-8093 Zürich, Switzerland**

Glycosyl phosphate glycosylating agents were successfully used in the synthesis of *C*-aryl linkages common to many natural products *via* a Lewis acid induced rearrangement. The rearrangement was stereo- and regio- specific, yielding only one *C*-glycoside product. *C*-alkyl glycoside carbohydrate mimetics were generated by using silicon derived *C* - nucleophiles and glycosyl phosphates. A short, high yielding synthesis of the *C*-glucoside 8,10-di-O-methylbergenin is described.

Introduction

C-glycoside natural products exhibit medicinally interesting properties, including antifungal and antitumorigenic responses (*1, 2*). Furthermore, *C*-glycosides have seen use as chiral building blocks in the synthesis of natural product macromolecules (*3-5*). *C*-glycoside analogs of biologically active carbohydrates are attractive pharmaceutical targets since they are not enzymatically degraded *in vivo* (*6-8*). Replacement of the exocyclic carbon-oxygen bond at the anomeric center with a carbon-carbon bond creates a hydrolytically stable carbohydrate mimetic.

Examination of specific carbohydrate-protein interactions can be accomplished with *C*-glycosides (Scheme 1). A series of *C*-glucosides and *C*-mannosides, such as **1**, were employed to study the binding differences between mannose and glucose specific lectins (*9*). *C*-Mannoside derivatives (**3**-**5**) were synthesized from *C*-allyl derivative **2** and used to block cell-surface lectins thereby inhibiting bacterial adhesion (*10*). The primary amine of **4** was functionalized with biotin to target proteins to the bacterial cell surface.

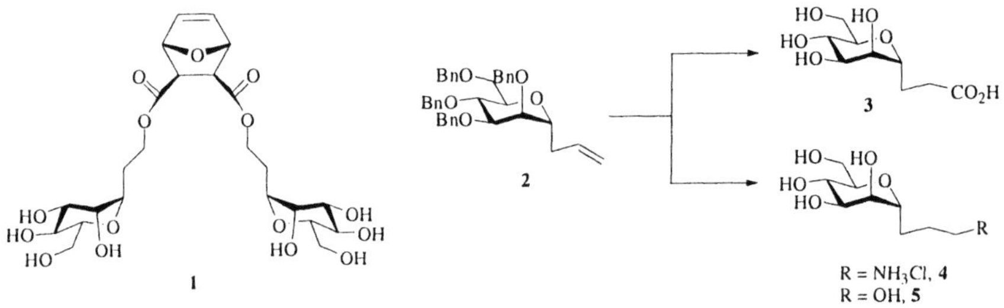

Scheme 1: C-glycosides used in biological assays.

The discovery of *C*-glycosidic natural products and the use of *C*-glycosides as carbohydrate mimetics have fueled investigations into the efficient synthesis of these compounds. Several different approaches for the synthesis of *C*-glycosides have been explored previously (*1, 2, 11*). The most common method for *C*-glycoside construction exploits the electrophilicity of the anomeric carbon by coupling nucleophiles to a glycosylating agent. Anomeric leaving groups used as glycosyl donors in this approach include anomeric acetates, trichloroacetimidates, thioglycosides, halides, and methyl glycosides (*1*, 2). Of interest to the work described here, anomeric phosphites (**6**, Scheme 2) were coupled to aromatic *C*-nucleophiles to yield *C*-aryl glycosides (*12*) and glycosyl phosphates were activated by SmI_2 to generate anomeric anions that were coupled with various electrophilic species (**10**, **13**, Scheme 2) (*13*). Other approaches to *C*-glycoside formation involve the use of transition metal anomeric complexes, anomeric anions, sigmatropic rearrangements, and palladium mediated couplings (*11, 14*).

An indirect route to *C*-aryl glycosidic linkages involves the initial installation of an *O*-glycosidic linkage followed by a *O*-to-*C* rearrangement (Scheme 3). A glycosyl donor is activated to generate an electrophilic anomeric species that couples to an aromatic phenol to afford an *O*-glycoside. The initial *O*-glycoside then rearranges to the *C*-aryl bond under acidic conditions (*15*). Various aromatic systems, such as naphthol, methoxyphenol and resorcinol derivatives, have been used in this approach. This rearrangement had been evaluated previously with glucosyl trichloroacetimidates (*16*) and fluorides (*17*) and exclusively afforded the sterically favored β-*C*-aryl glucoside product.

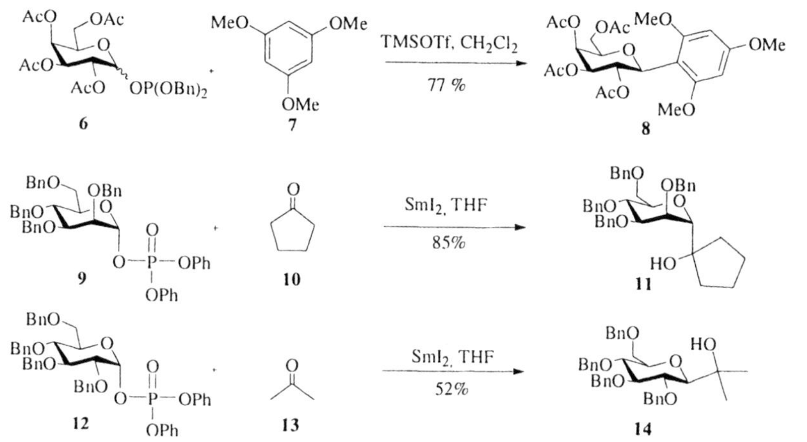

Scheme 2: Use of anomeric phosphorous compounds in C-glycoside formation.

The construction of *C*-alkyl glycosides can be achieved by the coupling of a carbon nucleophile and a glycosylating agent. The nucleophiles known to participate in *C*-glycosidations include organoaluminum, organolithium, organotin and Grignard reagents, while silicon based nucleophiles, such as trimethylsilyl enol ethers and allyltrimethylsilane, also have seen much use. The anomeric stereochemistry of *C*-alkyl glycoside formation is strongly dependent on the nature of the nucleophile, as was illustrated with the use of allyl and 2-methyl-2-propenyl trimethylsilane (**16** and **17,** Scheme 4). Generally, formation of the thermodynamically more stable α-glycosides predominates (*18*).

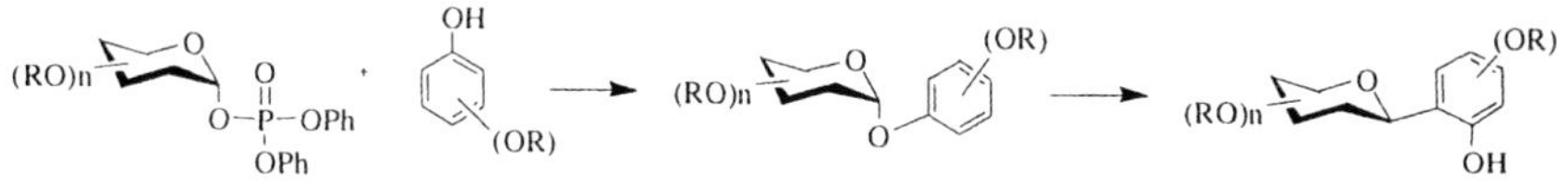

Scheme 3: O-to-C rearrangement.

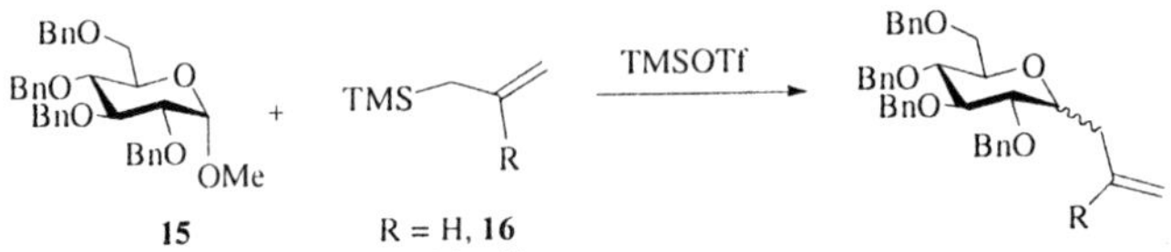

BnO BnO BnO BnO OMe + TMS R —TMSOTf→ BnO BnO BnO BnO R

15

R = H, **16**
R = Me, **17**

R = H, **18**
87% α:β = 100:0
R = Me, **19**
81% α:β = 89:11

Scheme 4: Anomeric selectivity in C-allyl formation.

Synthesis of *C*-Aryl Glycosides

Mannosyl donors had not previously been employed in the formation of aryl linkages via the *O*-to-*C* rearrangement pathway, thus the effect of the axial C2-substituent on the resulting stereochemistry was not known. To evaluate the rearrangement in the synthesis of *C*-aryl mannosides, phosphate **9** was prepared (*13*). Glycosyl phosphate **9** was activated with TMSOTf at 0 °C in the presence of electron-rich phenolic nucleophiles. Coupling of phosphate **9** with 3,4,5-trimethoxyphenol **20** afforded exclusively the α-*C*-linked glycoside **21** in just 30 minutes (Table 1). The rearrangement occurred even at low temperatures (-15 °C) and did not allow for the isolation of any *O*-glycoside product.

The aromatic phenol was varied to explore the scope of the *O*-to-*C* conversion with mannosyl phosphates. Using phosphate **9**, the α-*C*-mannosides of 2-naphthol and 3-benzyloxy phenol (**23** and **25**, Table 1) were synthesized in excellent yield. *O*-Mannosides were obtained exclusively with less nucleophilic aromatic systems, such as 3-acetoxy phenol. Several non-phenolic aromatic systems were unsuccessful in the formation of *C*-aryl or *O*-aryl glycosides. Reaction of **9** with furan, thiophene, trimethoxybenzene, and indole in the presence of TMSOTf did not result in any product formation. Interestingly, activation of **9** in the absence of any aromatic nucleophiles gave **26** as the major product via an intramolecular *C*-glycosylation (Figure 1) (*19*).

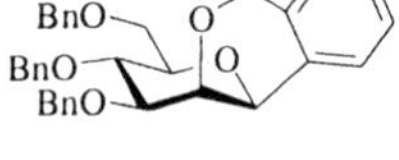

BnO BnO BnO O O

26

Figure 1. Intramolecular C-glycosylation.

Given the success with mannosyl phosphate **9**, the construction of *C*-aryl linkages via an *O*-to-*C* rearrangement with glucosyl phosphate **12** was investigated (*20*). Coupling of 3,4,5-trimethoxyphenol **20** and **12** afforded α-*O*-glycoside **27** (Figure 2) in 79% yield in 15 minutes. The rearrangement to the β-*C*-aryl linkage (**30**, Table 2) required higher temperatures and a longer reaction time (>3 h) due to the slower *O*-to-*C* conversion in the glucose series.

Table 1: C-Aryl mannosides.

Acceptor	*C-Glycoside product*	*Yield C-glycoside (α:β)*	*Yield O-glycoside (α:β)*
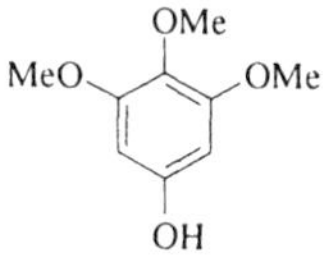 **20**	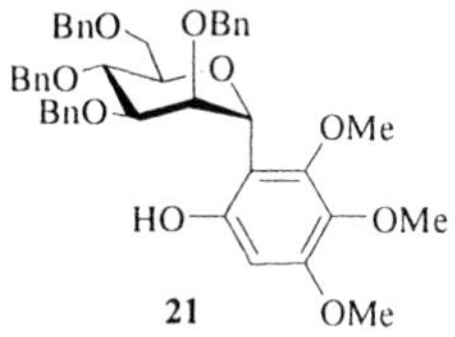 **21**	85% (1:0)	0%
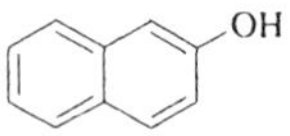 **22**	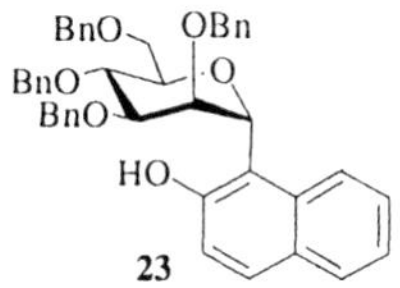 **23**	79% (1:0)	0%
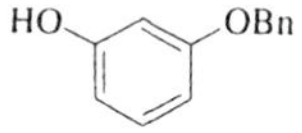 **24**	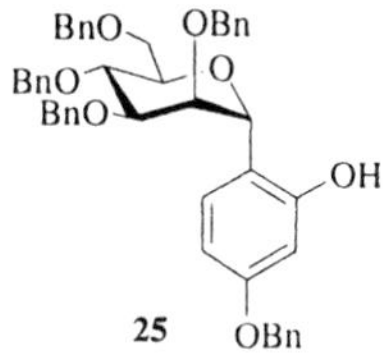 **25**	82% (1:0)	0%

In expanding this methodology to other phenols, 2-naphthol and 3-benzyloxy resorcinol were used as aromatic nucleophiles. The *O*-glycoside products (**28** and **29,** Figure 2) were isolated after 15 minutes at 0 °C, and *C*-glycosides (**31** and **32**) were obtained after longer reaction times. In all cases, the rearrangement was completely stereospecific, yielding exclusively the β-*C*-aryl linkage (Table 2). Interestingly, *O*-glycoside products were obtained in trace amounts even after extended reaction times (>3 h) indicating that the *O*-to-*C* conversion did not reach completion. Furthermore, *O*-glucosides were formed exclusively with less nucleophilic aromatic systems, as was observed with mannosyl phosphate **9**.

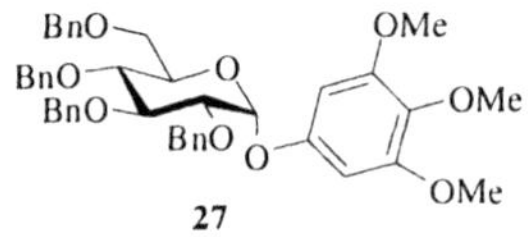

Figure 2. O-Glycosylation products.

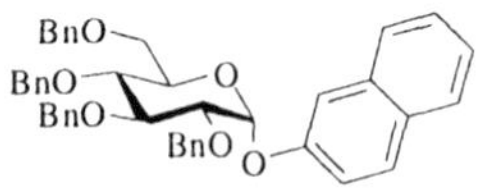

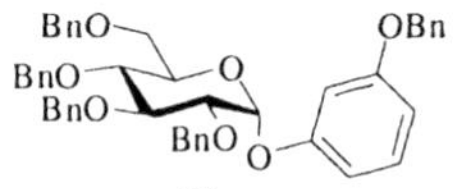

Table 2: C-Aryl glucosides

Acceptor	*C-Glycoside product*	*Yield C-glycoside (α:β)*	*Yield O-glycoside (α:β)*
20	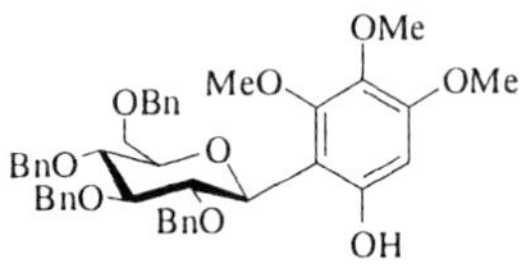 **30**	57% (0:1)	13% (1:0)
22	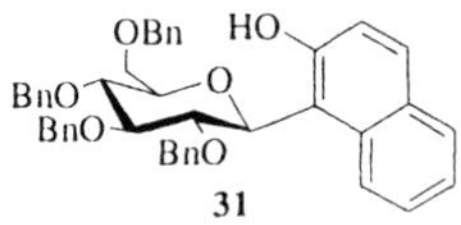 **31**	60% (0:1)	9% (1:0)
24	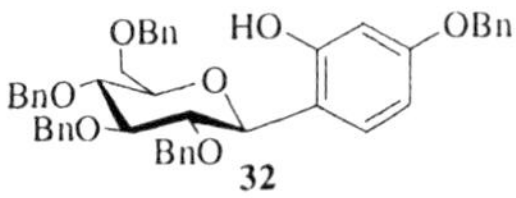 **32**	62% (0:1)	trace

Synthesis of *C*-Alkyl Glycosides

C-Alkyl glycosides are useful carbohydrate mimetics, commonly employed as enzyme inhibitors. The formation of various *C*-alkyl glycosides was evaluated with phosphates **9** and **12** (Table 3). Mannosyl phosphate **9** was activated with TMSOTf and coupled to allyltrimethylsilane to provide α-allyl glycoside **2** (*18*) in excellent yield (93%). Coupling of **9** to the cyclopentanone derived trimethylsilyl enol ether **35** afforded **34** as a mixture of diastereomers in 84%

yield (Table 3). Phosphate **12** was activated with TMSOTf at 0 °C in the presence of **16** to afford the α-allyl glycoside **33** (*21*). Glycosyl phosphate **12** was coupled to the silyl enol ether, resulting in the formation of **36** (*21*) as a mixture of diastereomers. Attempts to couple the glycosyl phosphates with trimethylsilyl cyanide and trimethylsilyl phenylacetylene did not afford any *C*-alkyl product.

Table 3: C-Alkyl glycosides.

Donor	*Acceptor*	*Product*	*Yield*
9	TMS (allyl) **16**	**2**	93%
12		OBn, BnO, BnO, O, BnO **33**	55%
9	OTMS **35**	OBn, OBn, BnO, BnO, O, O **34**	84%
12		OBn, BnO, BnO, O, BnO, O **36**	68%

Synthesis of 8,10-Di-*O*-Methylbergenin

Bergenin **37**, norbergenin **38** and 8,10-di-*O*-methylbergenin **39** (Figure 3) are gallic acid derived *C*-glycosides isolated from a variety of plants (*22, 23*). Numerous reports regarding the biological and pharmacological properties of bergenin-type *C*-glycosides have been disclosed. Bergenin enriched extracts from *Macaranga peltata* are used in Indian folk medicine for the treatment of venereal deseases (*24*), and acetylated bergenin has shown an antihepatotic effect in animal experiments (*12*). Bergenin itself shows activity against HIV and is the active pharmaceutical ingredient of a Chinese drug described to be effective against cough and bronchitis (*25*).

Despite the interesting biological properties of this class of natural products, few competent methods for the chemical synthesis of bergenin and its derivatives exist. Schmidt and coworkers reported a ten step synthesis of 8,10-di-*O*-

methylbergenin **39** in 8.8% overall yield from perbenzylated trifluoroacetyl glucose (*26*). Martin *et al.* developed a synthesis of **39** based on an intramolecular *C*-glycosylation of a 2-(3',4',5'-trimethoxy)benzyl *n*-pentenyl glucoside followed by oxidation of the benzylic methylene group. 8,10-Di-*O*-methylbergenin **39** was prepared in 12% yield over eight steps from peracetylated glucosyl bromide (*19*). Apart from the modest overall yield, these syntheses required numerous protecting group manipulations thus rendering them unsuitable for the synthesis of an extended set of bergenin derivatives.

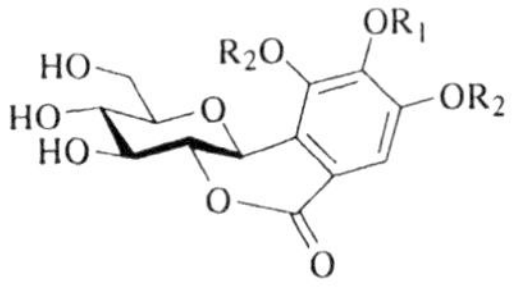

37 Bergenin $R_1 = CH_3$, $R_2 = H$
38 Norbergenin $R_1 = R_2 = H$
39 8,10-Di-*O*-methylbergenin $R_1 = R_2 = CH_3$

Figure 3. Bergenin structure.

Encouraged by the ease of access to *C*-aryl glycosides such as **30**, a synthesis of 8,10-di-*O*-methylbergenin was explored. While the effect of an electron withdrawing C2-protecting group on the *O*-to-*C* rearrangement had not been investigated, we envisioned the formation of a *C*-glycoside containing a temporary C2-ester would allow for the formation of the necessary lactone of the bergenin structures (Figure 3) (*27*). Phosphate **40** containing a C2-ester was synthesized from the corresponding glycal (Scheme 5) (*28*). Coupling of **40** to 3,4,5-trimethoxyphenol **20** was accomplished by TMSOTf activation to afford *C*-glycoside **41** (*29*) and *O*-glycoside **42** in 21% and 35% yield, respectively. The conversion required extended reaction times (12 h) to obtain the desired product in modest yield.

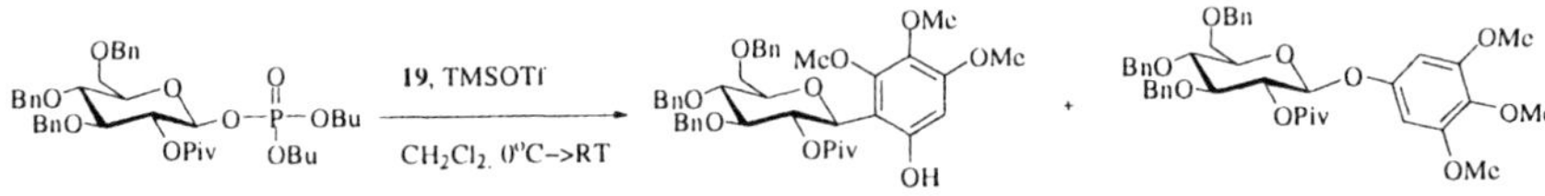

40 → (19, TMSOTf; CH_2Cl_2, 0°C->RT) → **41** (21%) + **42** (35%)

*Scheme 5: Formation of C-aryl glycoside **41**.*

The construction of 8,10-di-*O*-methylbergenin was attempted from the C2-benzyl *C*-glycoside **9** given the low yielding synthesis of **41**. It was postulated that lactone formation would be specific to the C2-hydroxyl and therefore the orthogonal protecting group was unnecessary. Treatment of the phenolic hydroxyl group of **30** with triflic anhydride/lutidine resulted in the formation of triflate **43** (Scheme 6).

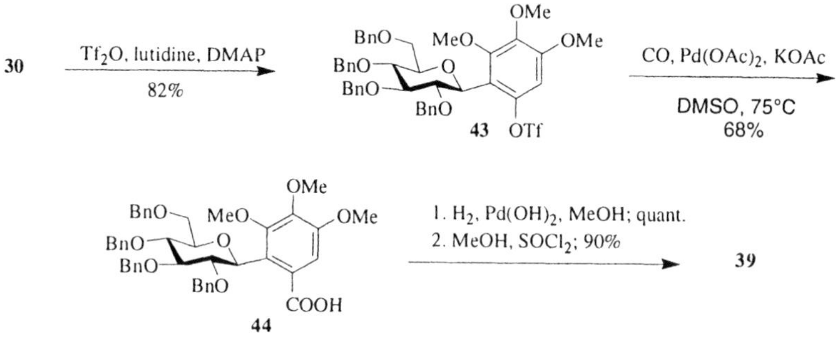

30 → Tf_2O, lutidine, DMAP; 82% → **43** → CO, $Pd(OAc)_2$, KOAc; DMSO, 75°C; 68% → **44** → 1. H_2, $Pd(OH)_2$, MeOH; quant. 2. MeOH, $SOCl_2$; 90% → **39**

Scheme 6: Completion of the bergenin synthesis.

Palladium (0) catalyzed carbonylation (*30*) at ambient pressure yielded 68% of *C*-glucosyl benzoic acid derivative **44**. Debenzylation of **44** by hydrogenation with Pearlman's catalyst in methanol provided the tetrahydroxyl *C*-glucoside in quantitative yield. Regioselective lactonization of the C2-hydroxyl group of the glucose scaffold was achieved by treatment of deprotected **44** with $SOCl_2$ in methanol to provide cleanly 8,10-di-*O*-methylbergenin **39** in an overall yield of 33% from **12**.

Summary

The efficient use of glycosyl phosphates in the syntheses of various *C*-aryl and *C*-alkyl glycosides has been demonstrated. Mannosyl and glucosyl phosphates were successfully employed in a Lewis acid induced *O*-to-*C* rearrangement and the construction of *C*-aryl linkages present in natural products was easily accomplished by this rearrangement. Use of silicon derived *C*-nucleophiles and glycosyl phosphates allowed for the generation of *C*-alkyl glycosides containing a functional handle for conjugation to proteins or molecular probes.

The utility of glycosyl phosphates in the synthesis of *C*-glycosides was demonstrated in the synthesis of the natural product 8,10-di-*O*-methylbergenin **39**. Utilizing the *O*-to-*C* rearrangement, the key *C*-aryl linkage was installed and further elaboration afforded the product in good yield.

Experimental

2-(2',3',4',6'-Tetra-O-benzyl-β-D-glucopyranosyl)-3,4,5-trimethoxy-phen-1-ol 30. Diphenyl 2,3,4,6-tetra-*O*-benzyl-α-D-glucosyl phosphate (25.0 mg, 0.032 mmol) was coevaporated with toluene, dissolved in CH_2Cl_2 (1.0 mL) and the solution was cooled to 0 °C. 3,4,5-Trimethoxy phenol (8.9 mg, 0.049 mmol) was added followed by the addition of TMSOTf (7.0 μL, 0.038 mmol). The reaction mixture was allowed to warm to ambient temperature over the following 4h. Et_3N (10 μL) was added and the solvent was removed *in vacuo*. Chromatography (15:1 Hexanes:EtOAc) afforded the product **30** as a coloress oil (13.0 mg, 57%) and the *O*-glycoside (3.0 mg, 13%). ^{1}H NMR (500 MHz) δ 7.54 (br s, 1H), 7.38 – 7.05 (m, 20H), 6.31 (s, 1H), 5.00 (d, *J* = 11.0 Hz, 1H), 4.90 – 4.85 (m, 3H), 4.58 (d, *J* = 11.9 Hz, 1H), 4.56 (d, *J* = 10.7 Hz, 1H), 4.45 (app t, *J* = 12.5 Hz, 2H), 4.09 (d, *J* = 10.7 Hz, 1H), 3.82 (t, *J* = 9.1, 1H), 3.86 – 3.74 (m, 12H), 3.71 – 3.68 (m, 1H), 3.61 – 3.58 (m, 1H). ^{13}C NMR (75 MHz) δ 154.3, 152.7, 151.9, 138.9, 138.1, 137.8, 135.7, 128.6, 126.6, 128.5, 128.4, 128.2, 128.0, 128.0, 128.0, 127.9, 127.9, 127.8, 127.8, 109.2, 97.5, 86.5, 80.8, 78.7, 75.8, 75.5, 75.3, 73.6, 68.2, 61.5, 61.0, 56.1. IR (thin film): 3369.1, 2858.9, 1611.9, 1494.2, 1363.4, 1075.6 cm^{-1}. $[\alpha]^{24}_D$: +7.0° (c 1.42, CH_2Cl_2). FAB MS *m/z* (M+H) calcd 729.3010 obsd 729.3034.

2-(2',3',4',6'-Tetra-O-benzyl-β-D-glucopyranosyl)-3,4,5-trimethoxy-benzoic acid 44. 2-(2',3',4',6'-Tetra-O-benzyl-β-D-glucopyranosyl)-1-trifluoro-methanesulfonyloxy-3,4,5-trimethoxybenzene **43** (100 mg, 0.119 mmol), potassium acetate (98 mg, 1.00 mmol), dppf (45 mg, 81 μmol), dppp (15 mg, 36 μmol) and 6.0 mg (25 μmol) palladium acetate were dissolved in DMSO (3 ml) in a flame dried flask. The solution was thoroughly degassed under high vacuum and purged with a stream of carbon monoxide for 10 min. The reaction was stirred under an atmosphere of carbon monoxide at 75 °C for 15 h, diluted with CH_2Cl_2 (60 ml) and extracted with 0.5 N HCl (20 ml). The organic phase was dried over $MgSO_4$ and coevaporated with toluene. Flash chromatography (100:30:1 hexane: EtOAc:AcOH) yielded the benzoic acid **44** as colorless, viscous oil (59.0 mg, 0.081 μmol).

^{1}H NMR (400 MHz, $CDCl_3$) δ 7.42 – 7.09 (m, 19H), 6.89 – 6.86 (m, 2H), 5.19 (d, *J* = 10.3 Hz, 1H), 5.00 – 4.87(m, 3H), 4.71 – 4.46 (m, 4H), 4.05 –

3.70 (m, 16H). ^{13}C NMR (100 MHz, $CDCl_3$) δ 168.9, 153.7, 152.9, 145.4, 138.9, 138.3, 138.0, 137.7, 129.5, 128.9, 128.6, 128.6, 128.5, 128.3, 128.2, 128.0, 127.9, 125.7, 122.9, 112.8, 87.1, 81.5, 79.9, 76.8, 76.1, 75.9, 75.8, 73.8, 67.6, 62.2, 61.3, 56.6. IR (thin film): 3030, 2939, 2867, 1725, 1590, 1496, 1453, 1360, 1118, 1088, 1068 cm^{-1}. $[\alpha]^{22}_D$: -6.9° (c 1, CH_2Cl_2). HR-ESI MS: $C_{44}H_{46}O_{10}$ *m/z* (M+Na) calcd 757.2983 obsd 757.2988.

References

1. Levy, D. E.; Tang, C. *The Chemistry of C-Glycosides.* Pergamon, **1995**.
2. Postema, M. H. D. *C-Glycoside Synthesis.* CRC Press, London, **1995**.
3. Smith III, A. B.; Zhuang, L.; Brook, C. S.; Boldi, A. M.; McBriar, M. D. *Tetrahedron Lett.* **1997**, *38*, 8667.
4. Lewis, M. D.; Cha, J. K.; Kishi, Y. *J. Am. Chem. Soc.* **1982**, *104*, 4976.
5. Aicher, T. D.; Buszek, K. R.; Fang, F. G.; Forsyth, C. J.; Jung, S. H.; Kishi, Y.; Scola, P. M. *Tetrahedron Lett.* **1992**, *33*, 1549.
6. Cheng, X.; Khan, N.; Mootoo, D. R. *J. Org. Chem.* **2000**, *65*, 2544.
7. Abe, H.; Shuto, S.; Matsuda, A. *Tetrahedron Lett.* **2000**, *41*, 2391.
8. Bazin, H. G.; Du, Y.; Polat, T.; Linhardt, R. J. *J. Org. Chem.* **1999**, *64*, 7254.
9. Weatherman, R. V.; Mortell, K. H.; Chervenak, M.; Kiessling, L. L.; Toone, E. J. *Biochemistry* **1996**, *35*, 3619.
10. Bertozzi, C. R.; Bednarski, M. D. *J. Org. Chem.* **1991**, *56*, 4326.
11. Jaramillo, C.; Knapp, S. *Synthesis* **1993**, 1.
12. Lin, C.-C.; Shimazaki, M.; Heck, M.-P.; Aoki, S.; Wang, R.; Kimura, T.; Ritzen, H.; Takayama, S.; Wu, S.-H.; Weitz-Schmidt, G.; Wong, C.-H. *J. Am. Chem. Soc.* **1996**, *118*, 6862.
13. Hung, S.-C.; Wong, C.-H. *Angew. Chem. Int. Ed.* **1996**, *35*, 2671.
14. Du, Y.; Linhardt, R. J. *Tetrahedron* **1998**, *54*, 9913.
15. Kometani, T.; Kondo, H.; Fujimori, Y. *Synthesis* **1988**, 1005.
16. Mahling, J.-A.; Schmidt, R. R. *Synthesis* **1993**, 325.
17. Matsumoto, T.; Hosoya, T.; Suzuki, K. *Synlett* **1991**, 709.
18. Hosomi, A.; Sakata, Y.; Sakurai, H. *Carbohydr. Res.* **1987**, *171*, 223.
19. Rousseau, C.; Martin, O. R. *Tetrahedron Assym.* **2000**, *11*, 409.
20. Hashimoto, S.; Honda, T.; Ikegami, S. *J. Chem. Soc., Chem. Comm.* **1989**, 685.
21. Hoffman, M. G.; Schmidt, R. R. *Liebigs Ann.* **1985**, 2403.
22. Saijo, R.; Nonaka, G.; Chen, I.-S.; Hwang, T.-H. *Chem. Pharm. Bull.* **1989**, 37, 2940.

23. Ramaiah, P. A.; Row, L. R.; Reddy, D. S.; Anjaneyulu, A. S. R.; Ward, R. S.; Pelter, A. *J. Chem. Soc., Perkin Trans 1*, **1979**, 2313.
24. Chopra, R. N.; Maya, S. L.; Chopra, I. C. *Glossary of Indian Medicinal Plants.* CSIR, New Delhi, **1956**.
25. Piacente, S. Pizza, C.; deTommasi, N.; Mahmood, J. *J. Nat. Prod.* **1996**, *49*, 565.
26. Frick, W.; Schmidt, R. R. *Carbohydr. Res.* **1991**, *209*, 101.
27. Hay, J. E.; Haynes, L. J. *J. Chem. Soc.*, **1958**, 2231.
28. Herzner, H.; Palmacci, E.; Seeberger, P. H. *Organic Letters* **1999**, *3*, 1547.
29. Schmidt, R. R.; Effenberger, G. *Carbohydr. Res.* **1992**, *171*, 59.
30. Cacchi, S.; Lupi, A. *Tetrahedron Lett.* **1992**, *33*, 3939.
31. Pyrek, J. S.; Kocor, M.; Sharma, B. R.; Atal, C. K. *Rocz. Chem.* **1977**, *51*, 1679.

Chapter 5

Aryl *C*-Glycosides by the Reverse Polarity Approach

Kathlyn A. Parker

Department of Chemistry, State University of New York at Stony Brook, Stony Brook, NY 11794-3400

Glycal-substituted quinols and quinol ketals, derived from the 1,2-addition of lithiated glycals to quinones or quinone monoketals, are versatile intermediates for the synthesis of aryl C-glycosides, providing access to all four substitution patterns found in the natural products. For example, reduction of a quinol provides a p-hydroxyaryl glycal (**8** → **9**, **11** → **12**). On the other hand, the Lewis acid-catalyzed isomerization of a quinol ketal gives a 2-hydroxy-5-methoxyphenyl glycal (**14** → **15**). Three-step procedures afford glycosylated phenols with additional substitution patterns. Thus, treatment of the silyl ether of a glycal-substituted quinol with a second lithiated glycal reagent followed by a Lewis acid leads to an o,p-diglycosylated phenol (**8a** → **18**, **19** → **21**). Reduction followed by Lewis acid treatment gives an o-glycosylated phenol (**24** → **30**). The substrates for the glycosylation schemes can be prepared regiospecifically. Thus the overall glycosylations can be carried out with regiocontrol. The practicality of the methodology is illustrated in total synthesis.

Introduction

The polyketide-derived aryl C-glycoside natural products have a variety of interesting biological activities, most notably antitumor activity (*1*). The defining feature of members of this class is a carbon-carbon bond linking a polycyclic aromatic "aglycone" and the anomeric carbon of a carbohydrate (Figure 1).

Figure 1

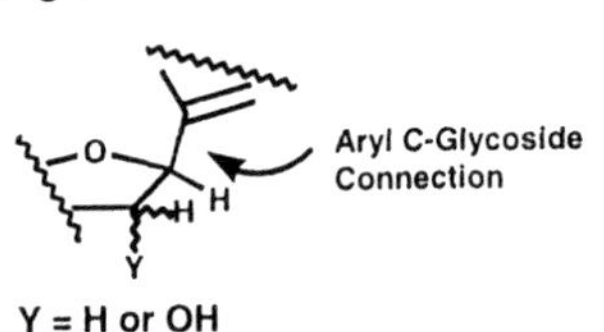

Y = H or OH

The first aryl C-glycoside was a synthetic compound, prepared as a nucleoside analog by Hurd and Bonner in 1945 (*2*). For some time there was intense interest in this class of compounds as antitumor agents, and more recently, as anti-HIV agents. Nevertheless, until recently, the only direct methods for the formation of aryl C-glycosides were based on variations of electrophilic aromatic substitution (*3*). These methods may afford mixtures of regioisomers and stereoisomers and were not generally amenable to the glycosylation of polycyclic aromatic "aglycones." With the discovery of the more synthetically challenging polyketide-derived natural products, several research groups initiated programs focused on the construction of the aryl C-glycoside linkage. As a result of this attention in the 1980's, improved methods based on the classical strategy and also methods based on new strategies have appeared (*4*). Furthermore, notable successes in the synthesis of aryl C-glycoside natural products have been achieved.

There is considerable variety in the structures of both the aglycone and the sugar moieties of these compounds. In particular, they may be α- or β-anomers, pyranosides or furanosides, 2'-deoxy or 2'-oxygenated, or amino glycosides. For the purposes of synthetic analysis, it is most useful to classify the naturally occurring aryl C-glycosides according to the substitution pattern of the glycosylated aromatic ring (*5*). Aside from the nogalamycins (which are C-5´ glycosides), they may be divided into four groups (Figure 2):

1) those in which the carbohydrate is attached *para* to a phenolic hydroxyl
2) those in which the carbohydrate is attached *ortho* to a phenolic hydroxyl
3) those in which two carbohydrates are attached *ortho* and *para* to a phenolic hydroxyl
4) those in which one carbohydrate is attached to a p-hydroquinone or quinone ring.

Figure 2. Naturally Occurring Polyketide C-Glycosides, Classified by Substitution Pattern

Group 1

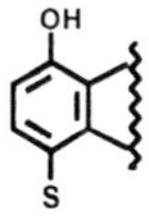

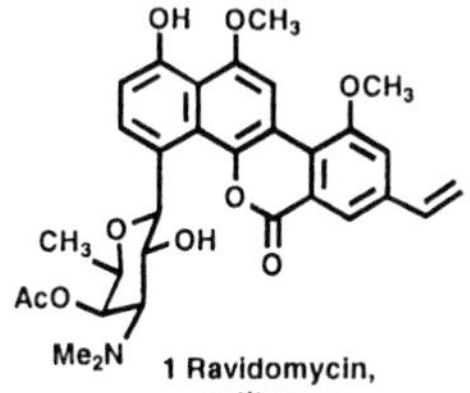

1 Ravidomycin, antitumor

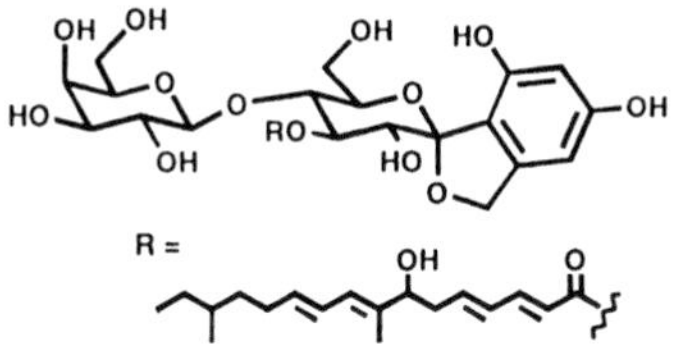

2 Papulacandin D, antifungal

Group 2

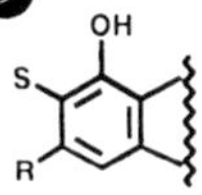

(R = H or CH_3)

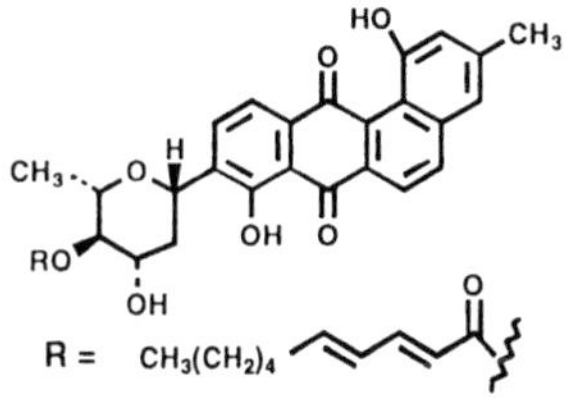

R = $CH_3(CH_2)_4$

3 C104 (L-glycoside) antibiotic

Group 3

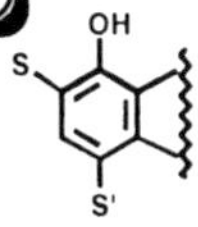

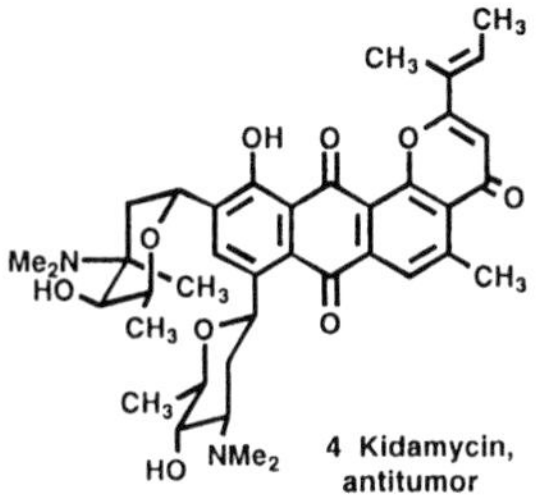

4 Kidamycin, antitumor

Group 4

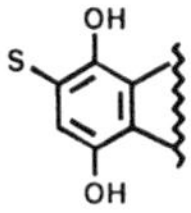

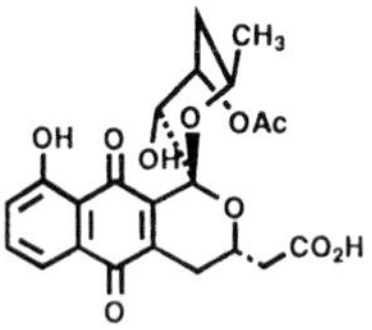

5 Griseusin B, antibiotic

Our interest in this area is in the recognition of unexploited reactivity patterns and the manipulation of this reactivity. Rather than optimize known approaches, we have chosen to seek opportunities for the development of new strategies and methods. As a result, we have been able to discover chemistry that leads to all four of the substitution patterns found in the naturally occurring aryl C-glycosides (*6*). The highlights of our explorations to date are summarized below.

Strategies and Method Development

Initially focusing our attention on targets with a group 1 substitution pattern, we settled on a "reverse polarity" or "umpolung" approach for the construction of the aryl C-glycoside bond. The standard methodology relied on electrophilic aromatic substitution with a C-1 halo sugar derivative acting as the electrophile. We decided to invert this polarity relationship and add a nucleophilic sugar equivalent (a lithiated glycal) to an electrophilic aromatic equivalent (a quinone or quinone derivative). Reduction of the quinol or quinol ketal and hydroboration of the glycal double bond would then afford the glycosylated phenol (*7*). A comparison of the two approaches is shown in Scheme 1.

Scheme 1.
The Classical and Umpolung Strategies in C-Aryl Glycoside Construction

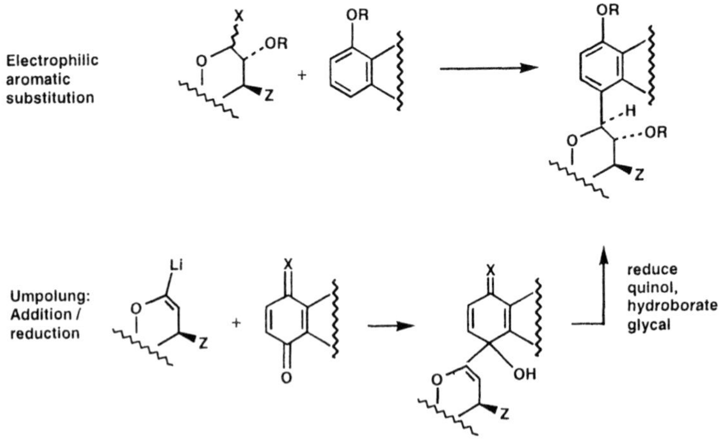

Electrophilic aromatic substitution

Umpolung: Addition / reduction

reduce quinol, hydroborate glycal

Lithiated glycals add to quinones to give quinols, the 1,2-addition products (e.g. **8**, **11**, Scheme 2). Reduction of benzoquinols to phenols (**8** -> **9**) was most conveniently effected with dithionite in the workup of the addition reaction. However, the dithionite conditions are apparently too acidic for naphthoquinol glycals which undergo fragmentation with this treatment. Naphthoquinols were conveniently reductively aromatized with aluminum amalgam (**10** → **11**) (*8*).

Scheme 2. Sequence for the Umpolung Strategy - Reduction to the Group 1 (Ravidomycin) Substitution Pattern

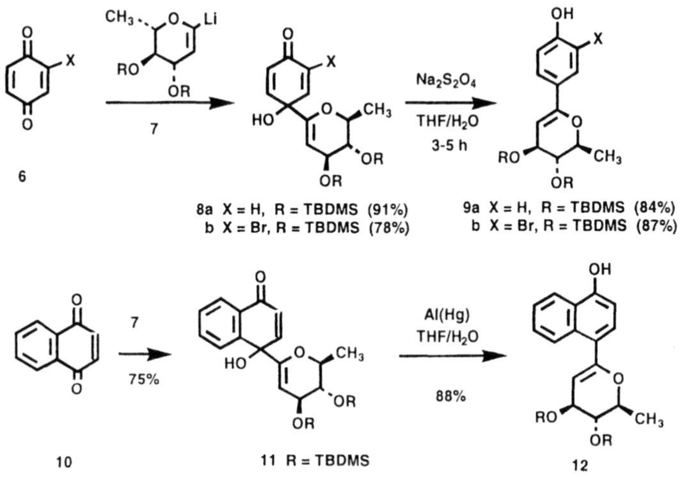

6 → (7) → 8a X = H, R = TBDMS (91%); b X = Br, R = TBDMS (78%) → ($Na_2S_2O_4$, THF/H_2O, 3-5 h) → 9a X = H, R = TBDMS (84%); b X = Br, R = TBDMS (87%)

10 → (7, 75%) → 11 R = TBDMS → (Al(Hg), THF/H_2O, 88%) → 12

An interesting feature of the chemistry in Scheme 2 is the regioselectivity of the glycosylation procedure with bromobenzoquinone. The directing effect of a halo substituent provides regiocontrol for this and other sequences (see below) that build on the addition of a lithiated glycal to a haloquinone.

Reductive aromatization of furanoglycal-substituted quinols give aryl C-glycals that are not stable. However, these compounds can be converted to aryl C-furanoglycosides by hydroboration (*9*).

While optimizing the conversion of quinols to phenols, we discovered that Lewis acidic hydride reagents effect the isomerization of glycal-substituted quinol ketal **14** to the ortho-substituted phenol **15** in a conversion that is competitive with reduction. Recognizing the attractiveness of this rearrangement as an avenue to aryl C-glycosides with the group 4 substitution pattern, we examined the reaction of ketal **14** with non-reducing Lewis acids. Even the mild conditions of exposure to zinc chloride in ether effected this dienone-phenol type rearrangement in quantitative yield (*10*).

Scheme 3. Preparation and Lewis Acid-catalyzed Rearrangement of Quinol Monoketals

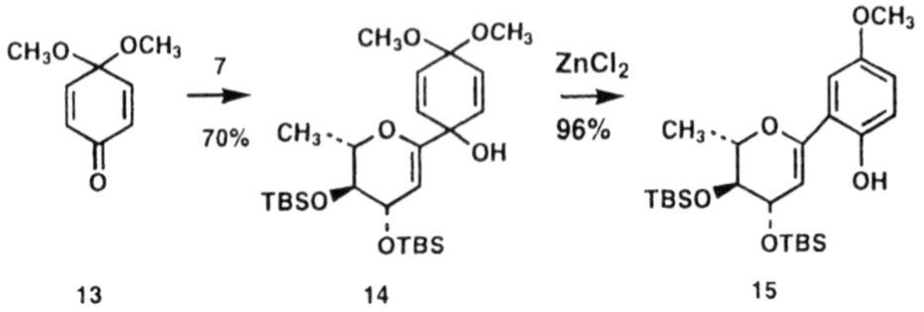

CH_3O OCH_3 → 7, 70% → CH_3O OCH_3, CH_3, O, OH, TBSO, OTBS → $ZnCl_2$, 96% → OCH_3, CH_3, O, OH, TBSO, OTBS

13 14 15

Noting that reduction of a quinol gave a product with the group 1 substitution pattern and that Lewis acid catalyzed rearrangement of a quinol ketal gave a product with the group 4 substitution pattern, we imagined that a combination of a reductive step and a rearrangement step might give products with the group 3 and possibly even the group 2 substitution patterns. The relationship of the methodology in hand and that suggested is shown in Scheme 4.

Scheme 4.
The Dienone / Phenol-type Rearrangement Suggests Additional Modifications.

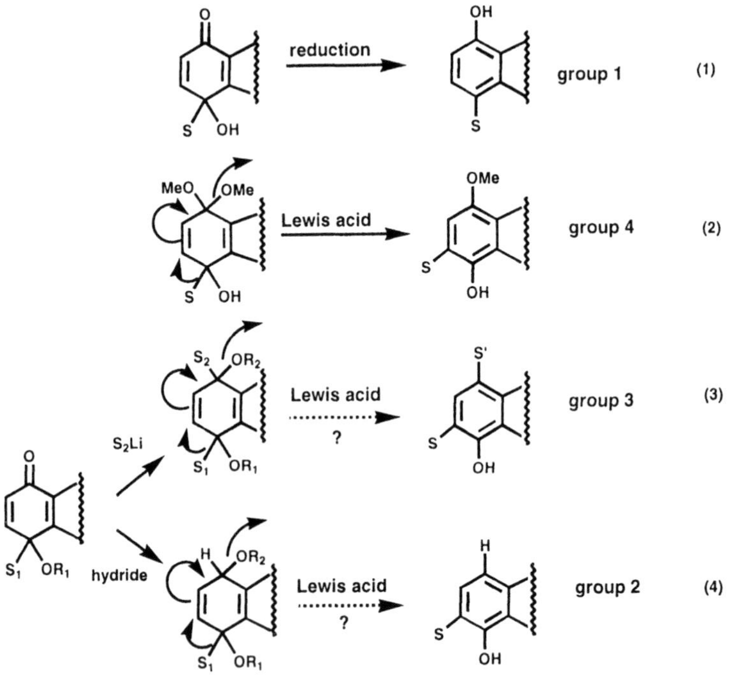

O, S, OH — reduction → OH, S — group 1 (1)

MeO, OMe, S, OH — Lewis acid → OMe, S, OH — group 4 (2)

S_2, OR_2, S_1, OR_1 — Lewis acid ? → S', S, OH — group 3 (3)

O, S_1, OR_1 — S_2Li / hydride

H, OR_2, S_1, OR_1 — Lewis acid ? → H, S, OH — group 2 (4)

Aside from questions of feasibility (i.e. would the less stable carbonium ions envisioned as intermediates in the third and fourth equations undergo the rearrangement steps?), there were other potential limitations of the proposed transformations. In the third equation, there are two apparently equivalent regiochemical possibilities. In the fourth equation, transannular loss of HOR_1 might be expected to compete with rearrangement.

In testing the feasibility of equation 3, we found that monosilyl ether substrates, prepared by lithiated glycal addition to a quinol silyl ether (Scheme 5), provided products that were the silyl ethers of phenols in which the first glycal group to be introduced was the one that migrated (*11*). Thus, the position of the silyl protecting group determines the regiochemistry of the rearrangement. Another way to look at this is that the order of addition of the two glycals determines which one of them becomes the substituent ortho to the phenolic hydroxyl group in the product.

Scheme 5. The TBS Glycal Rearrangement is Regiospecific. The Second Glycal to be Introduced is the Migrating Group.

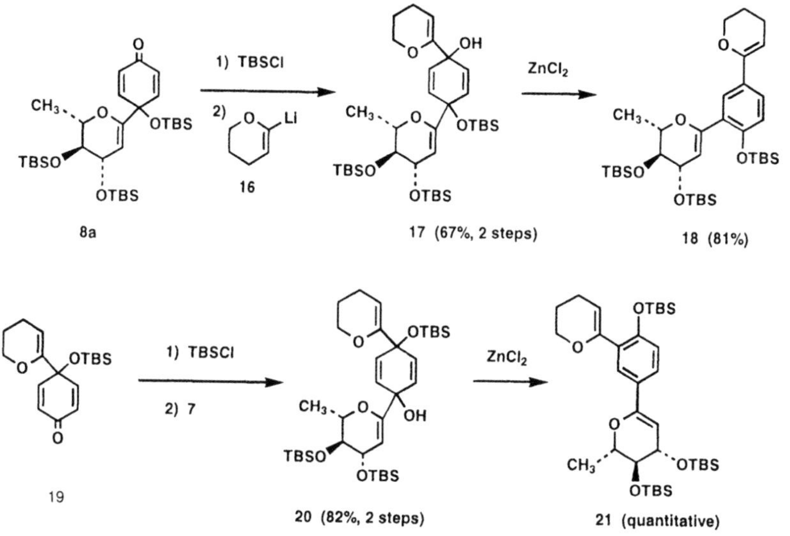

The newly discovered regiocontrolled construction of aryl bis glycals would be generally useful if the first glycal could be introduced regiospecifically into a complex ring system. For a solution to this problem, we relied on the known directing effect of halogen substituents on 1,2-additions to quinones. Thus, bromoquinone **22a** underwent addition with lithiated glycal **7** to give bromoquinol **23a** which was converted to its TES ether. Then a second lithiated glycal reagent was added (Scheme 6). Debromination of alcohol **25a** was effected by transmetallation and the resulting rearrangement substrate **26a** was treated with zinc chloride in ether to afford the bis glycal **27a**. The isomeric bromonaphthoquinone **22b** was subjected to the same series of reagents to give, after the five-step sequence, the isomeric naphthol **27b** (*12*).

Scheme 6. Regiocontrol in the Preparation of the Aryl C-Glycal Relies on the Availability of a Regiochemically Clean Substrate

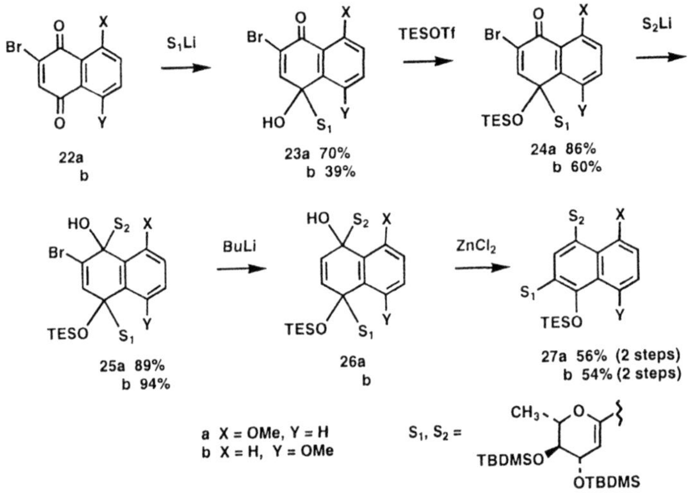

22a, b → (S_1Li) → 23a 70%, b 39% → (TESOTf) → 24a 86%, b 60% → (S_2Li) → 25a 89%, b 94% → (BuLi) → 26a, b → ($ZnCl_2$) → 27a 56% (2 steps), b 54% (2 steps)

a X = OMe, Y = H
b X = H, Y = OMe

S_1, S_2 = (glycal: CH_3, O, TBDMSO, OTBDMS)

The final test of the versatility of the reverse polarity approach would be that of completing equation 4. Remarkably, the TBS group served here also to promote the desired rearrangement. Thus, bromoquinol **24a** was reduced with dibal to the alcohol **28a** (Scheme 7).

Scheme 7. Regiocontrol in the Preparation of the Group 2 Aryl C-Glycals

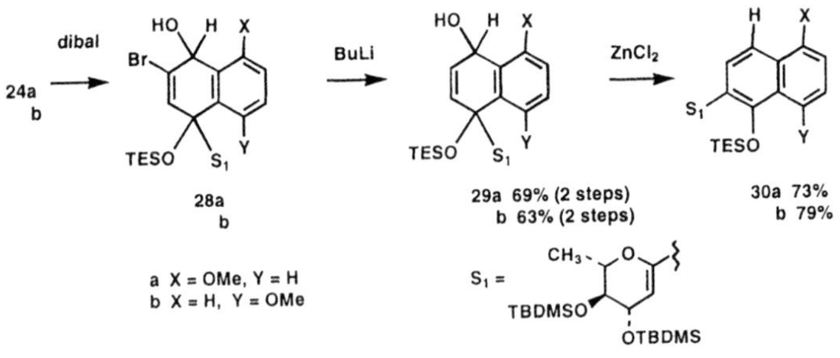

24a, b → (dibal) → 28a, b → (BuLi) → 29a 69% (2 steps), b 63% (2 steps) → ($ZnCl_2$) → 30a 73%, b 79%

a X = OMe, Y = H
b X = H, Y = OMe

S_1 = (glycal: CH_3, O, TBDMSO, OTBDMS)

Debromination gave the rearrangement substrate **29a** and the rearrangement proceeded smoothly to afford naphthol **30a**. On the other hand, the isomeric bromoquinol **28b**, after a similar sequence, afforded the isomeric naphthol **30b** (*12*).

Applications in Total Synthesis

Seeking to demonstrate the merits of the reverse polarity approach to aryl C-glycosides in a total synthesis, we have tested it in the context of the antifungal papulacandin / chaetiacandin class (group 1 substitution pattern) and also in the context of the antibiotic C104 (group 2 substitution pattern). These applications are shown below (Scheme 8).

Scheme 8. Synthesis of the Papulacandin and Chaetiacandin Nuclei

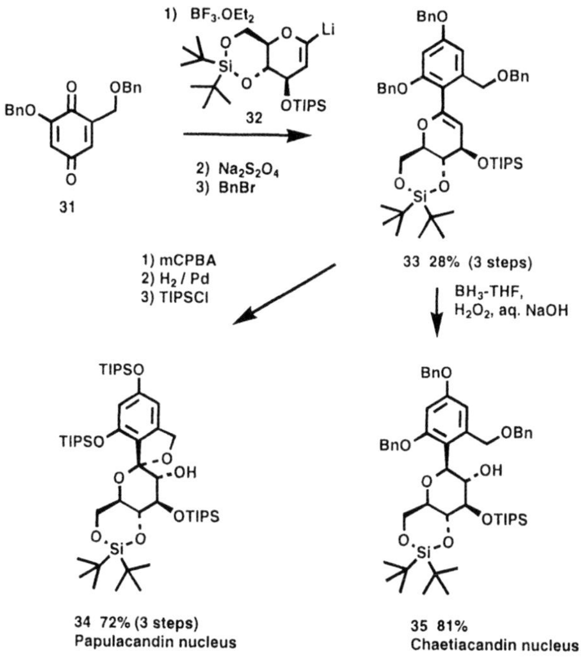

31

1) $BF_3.OEt_2$, 32

2) $Na_2S_2O_4$

3) BnBr

33 28% (3 steps)

1) mCPBA

2) H_2 / Pd

3) TIPSCl

BH_3-THF, H_2O_2, aq. NaOH

34 72% (3 steps)
Papulacandin nucleus

35 81%
Chaetiacandin nucleus

The papulacandin nucleus was generated by elaboration of quinone **31**. Thus addition of the lithiated glycal **32** to quinone **31** followed by dithionite reduction of the quinol adduct and benzylation gave aryl C-glycal **33**. Peracid oxidation, debenzylation with ketal formation, and silylation converted the aryl C-glycal **33** to the spiro glycoside **34** in which the papulacandin nucleus is appropriately protected for elaboration to the more complex natural products (see **2**). On the other hand, hydroboration converted glycal **33** directly to glycoside **35**. This structure represents the nucleus of chaetiacandins, antifungal antibiotics related to the papulacandins. Again the nucleus is selectively protected for elaboration to members of the antibiotic series (*13*)

The enantiomer of the antibiotic C104 was prepared by a route based on the five step reduction / migration sequence that was developed for compounds with the group 2 substitution pattern. Thus, the quinone "aglycone" **36** was converted to the aryl C-glycal **40** (Scheme 9). Reduction of the glycal double bond was accompanied by selective removal of the TES group from the sugar moiety. Introduction of the decadienoic acid sidechain followed by global deprotection and air oxidation gave **ent-C104** (*14*).

Scheme 9. The Five-step Glycosylation Sequence in the Synthesis of ent- C104

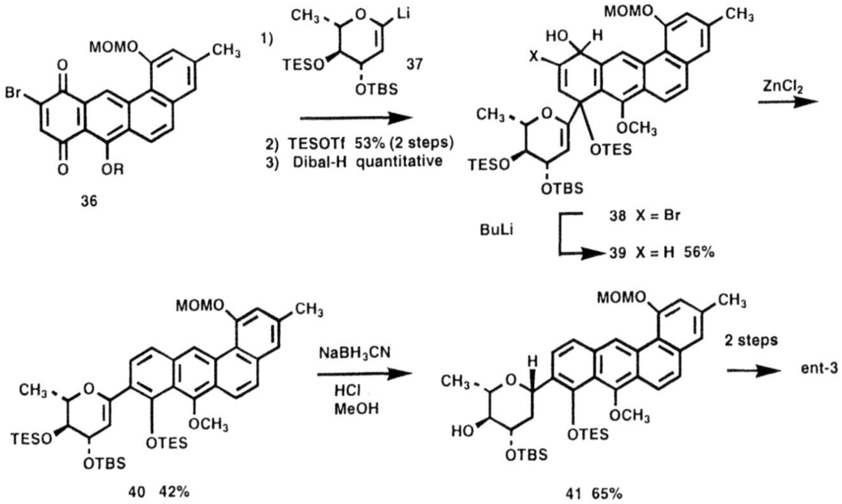

Summary

Investigations of the reverse polarity (umpolung) approach to the preparation of aryl C-glycosides have revealed the novel chemistry of glycal-substituted quinols. Reactions discovered during the course of these studies serve as the basis for short schemes that establish the functional group patterns required for the preparation of members of all four structural classes of the aryl C-glycoside natural products.

Experimental

Aryl C-Glycal 33. To a solution of fully protected glycal (corresponding to lithiated reagent 32, 180 mg, 0.407 mmol) in THF (0.15mL) at -78 oC (dry ice/acetone bath) was added t-BuLi (1.7M in hexanes, 0.48 mL, 0.82 mmol). The reaction mixture immediately turned bright yellow. After 5 min of stirring at -78 °C, the reaction mixture was allowed to warm to 0 °C (ice bath) and was then stirred for an additional 105 min. The resulting solution, having now turned a pale yellow color, was cooled to -100 °C (liq N_2/pentanes bath) and was added via cannula to a solution of quinone **31** (75 mg, 0.225 mmol) and BF_3-Et_2O (0.03 mL, 0.225 mmol) in THF (1.5 mL) at -78 °C. The reaction mixture was stirred at -78 °C for another 8 h, quenched with water, and allowed to warm to ambient temperature. The reaction mixture was extracted with ether (3 x 25 mL) and the resulting organic solution was washed with water (3 x 25 mL) and brine (1 x 25 mL) and then dried over Na_2SO_4. Column chromatography (3:1 Hex/EtOAc with 0.1% Et_3N) afforded 59 mg (33%) of quinol as a pale yellow syrup and 46 mg (61%) of recovered quinone **31**. ^{1}H NMR ($CDCl_3$) δ 7.41-7.25 (m, 10 H), 6.54 (s, 1 H), 6.36 (s, 1 H), 5.84 (bs, 1 H), 5.10 (s, 2 H), 4.93 (d, $J_{2,3}$ = 2.4 Hz, H-2), 4.72 and 4.60 (2d, J = 12.0 Hz, AB, 2 H), 4.55 (d, J = 1.6 Hz, 2 H), 4.46 (dd, $J_{2,3}$ = 2.4 Hz, $J_{3,4}$ = 6.8 Hz, H-3), 4.14-4.09 (m, H-6eq), 4.06 (dd, $J_{4,5}$ = 10.0 Hz, $J_{3,4}$ = 6.8 Hz, H-4), 3.96 (dd, $J_{5,6ax}$ = 10.0 Hz, $J_{6eq,6ax}$ = 10.0 Hz, H-6ax.), 3.85 (ddd, $J_{4,5}$ = 10.0 Hz, $J_{5,6ax}$ = 10.0 Hz, $J_{5,6eq}$ = 4.8 Hz, H-5), 1.13-0.99 (m, 39 H). IR ($CDCl_3$) 3424, 2942, 2864, 1658 cm^{-1}

To a solution of quinol (200 mg, 0.026 mmol) in THF/H_2O (5:2, 0.7 mL) was added $Na_2S_2O_4$ (360 mg, 0.206 mmol). The reaction mixture was stirred at room temperature for 10 h and then concentrated. The resulting residue was dissolved in ether (5 mL) and this solution was washed with water (2 x 5 mL) and then dried over Na_2SO_4. Solvent was removed under reduced pressure and crude phenol was redissolved in THF (1 mL). This solution was cooled to -78 °C, NaH (2 mg, 0.08 mmol) was added, and the reaction mixture was warmed to ambient temperature. Benzyl bromide (44 mg, 0.026 mmol) was added and the reaction mixture was stirred overnight and then quenched with cold water (3 mL). Extraction with $CHCl_3$ (3 x 15 mL) gave a combined organic solution that was washed with water (3 x 15 mL) and brine (1 x 15 mL) and then dried over $MgSO_4$. Column chromatography (6:1 Hex/EtOAc) gave 23 mg (85%) of a colorless syrup.

^{1}H NMR ($CDCl_3$) δ 7.43-7.26 (m, 15 H), 6.79 (d, J=3.0 Hz, 1 H), 6.53 (d, J = 3.0 Hz, 1H), 5.05 (s, 2 H), 5.02 (s, 2 H), 4.74 (d, $J_{2,3}$ = 2.1 Hz, H-2), 4.60 (d, J = 3.0 Hz, 2 H), 4.54 (d, J = 3.0 Hz, 2 H), 4.50 (dd, $J_{2,3}$ = 2.1 Hz, $J_{3,4}$ = 6.0 Hz, H-3), 4.16-4.01 (m, H-6eq., H-4), 3.92 (dd, $J_{5,6ax}$ =10.2 Hz, $J_{6eq,6ax}$ = 10.2 Hz, H-6ax), 3.89 (ddd, $J_{4,5}$ = 10.2 Hz, $J_{5,6ax}$ =10.2 Hz, $J_{5,6eq}$ = 4.2 Hz, H-5), 1.15-0.88 (m, 39 H). ^{13}C NMR ($CDCl_3$) δ 160.20, 158.0, 147.1, 139.9, 138.3, 137.1, 136.8, 128.6, 128.4, 128.3, 128.0, 127.7, 127.6, 127.5, 126.9, 117.3, 107.0, 105.1, 100.2, 73.1, 72.3, 71.9, 70.4, 70.1, 69.4, 66.1, 27.5, 27.0, 22.7, 19.9, 18.2, 12.5.
IR ($CDCl_3$) 2938, 2863 cm^{-1}
HRMS (FAB) calc. for $C_{51}H_{70}O_7Si_2Na$ ($[M+Na]^+$): 873.4558. Found: 873.4583.

Chaetiacandin nucleus **35**. To a solution of aryl C-glycal **33** (5 mg, 6 μmol) in THF (0.1 mL) at 0 °C was added BH_3-THF (1M in THF, 8.0 eq., 50 μL, 0.05 mmol). The reaction mixture was stirred at ambient temperature for 24 h and then treated with a solution of 3 N NaOH/30% H_2O_2 (1:1, 1.2 mL). The reaction mixture was stirred for 24 h then diluted with CH_2Cl_2, washed with aqueous 20% sodium hydrogen sulfite (1 x 10 mL), saturated aqueous ammonium chloride (1 x 10 mL), water (1 x 10 mL) and brine (1 x 10 mL) and then dried over Na_2SO_4. Solvent was evaporated and column chromatography (3:1 Hex/EtOAc) of the residue afforded 4 mg (approximately 80%) of a colorless syrup. ^{1}H NMR [$(CD_3)_2SO$, 140 °C] δ 7.48-7.24 (m, 15 H), 6.76 (d, J= 2.4 Hz, 1 H), 6.74 (d, J= 2.4 Hz, 1 H), 5.12 (s, 2 H), 5.05 (s, 2 H), 4.74 and 4.55 (2d, J= 12.4 Hz, AB, 2 H), 4.67 (d, $J_{1,2}$ =10.0 Hz, H-1), 4.54 (s, 2 H), 4.02-3.96 (dd, centered at 4.00, $J_{1,2}$ =10.0 Hz, $J_{2,3}$ =5.2 Hz, H-2, overlapping m, 3.99, H-3), 3.69 (t, J = 8.0 Hz, H-6, coupled to H-5 and to geminal H-6 with the same coupling constant), 3.63 (m, $J_{4,5}$ = 10.0 Hz, H-4), 3.62 (t, J = 8.0 Hz, H-6, coupled to H-5 and to geminal H-6 with the same coupling constant), 3.38 (m, H-5), 1.23-0.87 (s, 39 H).

IR ($CDCl_3$) 3453, 2938, 2862 cm^{-1}
HRMS (FAB) calc for: $C_{51}H_{72}O_8Si_2Na$ ($[M+Na]^+$): 891.4664. Found: 891.4645.

Acknowledgment

The work described in this chapter was carried out by my very capable and hard-working coworkers who were graduate students at Brown University and whose contributions are acknowledged individually in the references. Their skillful laboratory techniques and careful observations have been essential for the success of this project. Our efforts have been supported by the National Institutes of Health (CA50720) and by the National Science Foundation (CHE-9521055). During part of the period during which this work was carried out, KAP held an NSF Career Advancement Award.

References

1. For an excellent general review, see Hacksell, U.; Daves, G. D., Jr. *Prog. Med. Chem.* **1985**, *22*, 1.
2. Hurd, C. D.; Bonner, W. A. *J. Am. Chem. Soc.* **1945**, *67*, 1972.
3. Hanessian, S. Pernet, A. G. *Adv. Carbohydr. Chem. Biochem.* **1976,** *33,* 111.
4. (a) For a comprehensive review of methods which establish C-glycoside bonds, see Postema, M. H. D. *Tetrahedron,* **1992,** *48,* 8545. Also, see (b) Suzuki, K. Matsumoto, T. "Total Synthesis of Aryl C-Glycoside Antibiotics" in *Recent Progress in the Chemical Synthesis of Antibiotics and Related Microbial Products*; Lukacs, G., Ed.; Springer: Berlin, 1993, Vol. 2, pp 353. (c) Jaramillo, C. Knapp, S. *Synthesis*, **1994**, 1. (d) Postema, M. H. D. *C-Glycoside Synthesis*, CRC Press, Ann Arbor, 1995.
5. Parker, K. A. *Pure Appl.Chem.* **1994**, *66*, 2135.
6. For an outline of an alternative "unified strategy" for the synthesis of aryl C-glycosides, see Martin, S. F. *Pure Appl. Chem.* **2003**, *75*, 63.
7. Proof of principle for the strategy was obtained in the ketal series; see Parker, K.A.; Coburn, C. A. *J. Am. Chem. Soc.* **1991**, *113*, 8516.
8. Parker, K. A.; Coburn, C. A.; Johnson, P. D.; Aristoff, P. *J. Org. Chem.* **1992**, *57*, 5547.
9. Parker, K. A.; Su, D.-S. *J. Org. Chem.* **1996**, *61*, 2191.
10. Parker, K. A.; Coburn, C. A.; Koh, Y.-h. *J. Org. Chem.* **1995**, *60*, 2938.
11. Parker, K. A.; Koh, Y.-h. *J. Am. Chem. Soc.* **1994**, *116*, 11149.
12. Parker, K. A.; Koh, Y.-h. Unpublished results.
13. Parker, K. A.; Georges, A. T. *Org. Lett.* **2000**, *2*, 497.
14. Parker, K. A.; Park, K.-J.; Lee, M. D. Unpublished results.

Chapter 6

The Ramberg-Bäcklund Way to *C*-Glycosides

Peter S. Belica, Alain Berthold, Nikki Charles, Guangwu Chen, Ajit Parhi, Paolo Pasetto, Jun Pu, Cuangli Yang, and Richard W. Franck*

Department of Chemistry, Hunter College, 695 Park Avenue, New York City, NY 10021

The invention of a four-step convergent process for the synthesis of C-glycosides is described. A carbohydrate and an aglycone, first linked by sulfur, then undergo the title reaction to forge a C-C double bond between the sugar and the aglycone. The key double bond is converted to a single bond. Tables of products are presented; additionally, examples of failed conversions are given.

Introduction

The Ramberg-Bäcklund reaction, a novel carbon-carbon bond forming transformation, was discovered in 1940 when ethyl chloroethylsulfone **1**, upon treatment with NaOH, was shown to form a mixture of cis and trans 2-butene. (Fig. 1) (*1*).The reaction has been used sporadically over the intervening years and has just received a comprehensive review in 2003.(*2-8*)

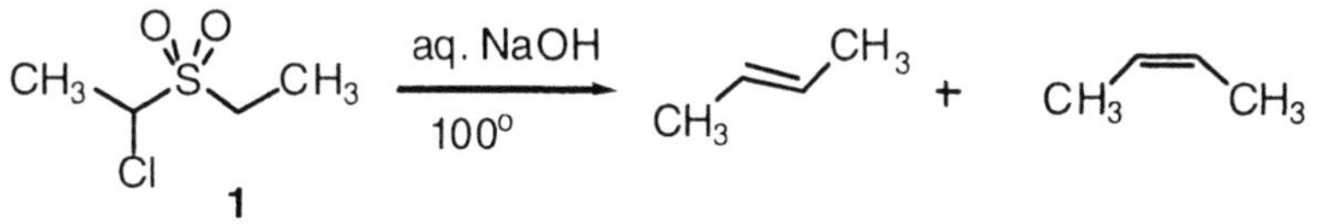

CH_3 **1** (Cl, SO_2) → aq. NaOH, 100° → trans $CH_3CH{=}CHCH_3$ + cis $CH_3CH{=}CHCH_3$

Fig. 1

In 1997, two of us (P.S.B. and R.W.F.) proposed its use for preparing C-glycosides from thioglycosides as part of our efforts to prepare the C-glycoside analog **3** of daunomycin **2**, a target that had been the focus of the Acton group in the 1980's (Fig. 2).(*9, 10*)

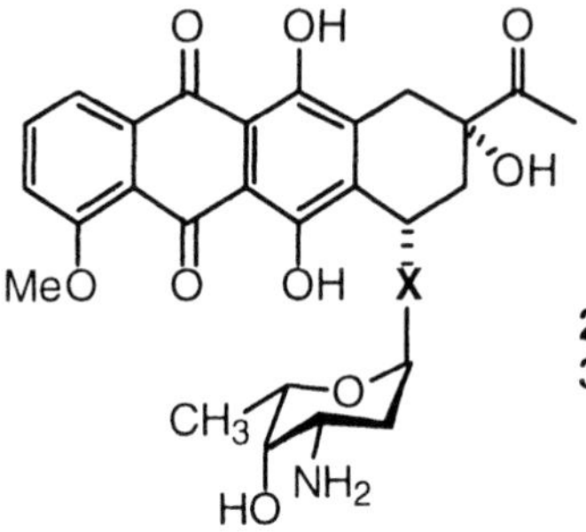
O
OH
O
OH
MeO
O
OH
X
CH3
O
HO
NH2

2 X = O daunomycin
3 X = CH_2 C-analog

Fig. 2

A literature survey in 1997 unearthed no example of this application of the RB reaction, although Hart had used the method to form an acyclic precursor to an aryl C-glycoside.(*11*) Therefore, an exploration of the reaction sequence illustrated in Scheme 1 was undertaken: (i) formation of S-glycoside **5**; (ii) sulfone glycoside **6**; (iii) Ramberg-Bäcklund to form exo glycal **10**; (iv) hydrogenation to afford C-glycoside **11**. This chapter will describe our successful development of this reaction sequence.(*12-22*)

Scheme 1 **The Ramberg-Backlund Reaction Applied to C-Glycoside Synthesis**

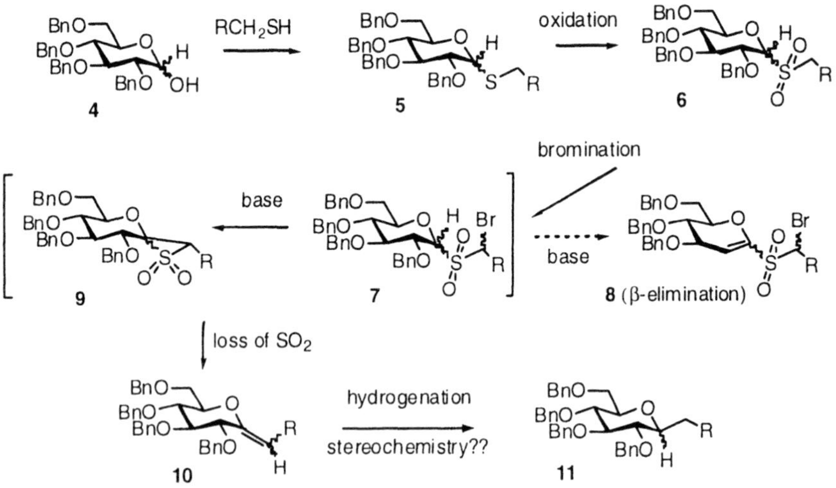
RCH2SH
oxidation
bromination
base
base
8 (β-elimination)
loss of SO2
hydrogenation
stereochemistry??
4
5
6
7
9
10
11

The conversion of carbohydrates to their thioglycoside **5** and sulfone derivatives **6** is routine and has been performed thousands of times by many groups using a variety of methods. The step in the overall process that was the focus of our concern was the base-catalyzed cycloelimination of brominated sulfone **7** to episulfone **9**. Thus, the β-elimination of a C-2 oxygen substituent to form a glycal species **8** was considered a possibility. In fact, our first two examples , entries 1 and 2 in Table 1, were chosen from the 2-deoxyglucose series just to avoid the β-elimination step. In the event, the unwanted elimination was rarely observed, even in the examples of mannose (entry 13, Table 2) and altrose (entry 11, Table 2) which have a geometry appropriate for a concerted anti elimination. Although there are described many permutations of base and halogen source for the halogenation and subsequent cycloelimination, we chose to use the Chan variation.(*23*) This entails a heterogeneous mixture of KOH supported on alumina and the innocuous bromine source, CF_2Br_2 (b.p. 23°) with a solvent mixture of CH_2Cl_2/t-BuOH. The advantage of this "one-pot" procedure for the two steps is that the excess base is removed by filtration, and after evaporation of solvent, the residue is usually a very clean mixture of product and unreacted starting material. The disadvantage is that the reaction cannot be heated. Thus, in the non-benzylic sulfone examples of Table 3, we often used $C_2F_4Br_2$ (b.p. 47°) as the bromine source so reactions could be run under "forcing" conditions of refluxing CH_2Cl_2 (entries 16b,18,19,20). The obstacle to the use of this higher-boiling freon is that it has been placed on the EPA "ozone" list and is no longer available in the U.S. It should be noted that Taylor reports the use of KO-t-Bu or KOH/CCl_4 for difficult cases, but our experience is that Ramberg-Bäcklund reactions are not as clean or high-yielding with this reagent (entry 17b). Entries 15-17 are worthy of note because these galactose cases are the only examples where significant amounts of bromovinyl products are obtained. This is an indication that the sulfones underwent a second bromination before the cycloelimination of HBr to produce the Ramberg-Bäcklund product. We do not have a rationalization for the remote effect of the 4-axial oxygen of galactose on the events at C-1 and C-1'. A complete experimental for each of the useful freons will be found at the conclusion of this chapter.

For the sake of completeness, we wish to describe some of our significant failures to obtain high yields of Ramberg-Bäcklund products from carbohydrate sulfones. An interesting example is the α-configured sulfone **12a** related to the β-sulfone **12b** of entry 18. This α-configured material is essentially recovered unchanged from the Ramberg-Bäcklund reaction mixture. A control experiment showed that KOH is not sufficiently basic to epimerize the α-configured sulfone to the β isomer which behaves well. We speculate that the conformation where the anomeric C-H bond bisects the OSO angle is not an accessible energy minimum for **12a** (e.g. conformer a, Scheme 2). On the other hand, the β-sulfone's α-H can be bisected by the OSO of the sulfone (conformer b). Such bisecting conformations are considered to be most favorable for acidification by the sulfone.(*24*) The α'-C-H bonds in both the α and β-glycosides are bisectable, but must not be sufficiently acidic. This α-epimer non- reactivity is not observed in the benzylic cases because the necessary α'-C-H reactivities are enhanced by their conjugation to the aromatic ring.

Table 1 Examples of Ramberg-Backlund reactions of benzylic glucosyl sulfones

	sulfone	Ramberg-Backlund product	yield, E/Z ratioi, notes
1	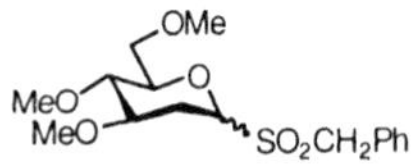 OMe, MeO, MeO, O, SO_2CH_2Ph	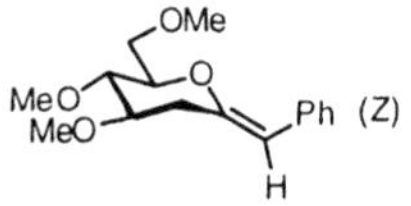 OMe, MeO, MeO, O, Ph (Z), H	72% Z/E= 83/ 17 also 9% endocyclic glycal
2	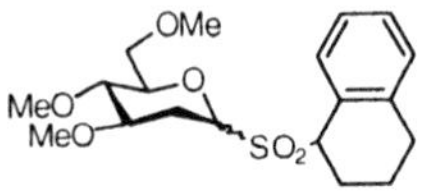 OMe, MeO, MeO, O, SO_2	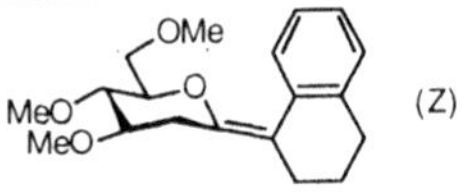 OMe, MeO, MeO, O, (Z)	76% Z/E=45/55
3	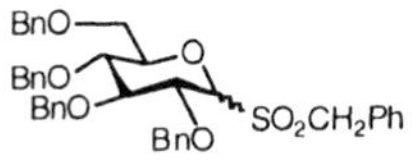 BnO, BnO, BnO, O, BnO, SO_2CH_2Ph	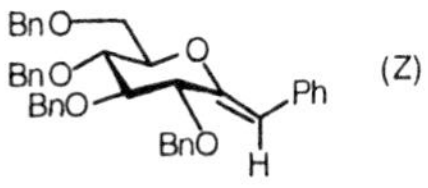 BnO, BnO, BnO, O, BnO, Ph, H (Z)	85% Z/E=91/9
4	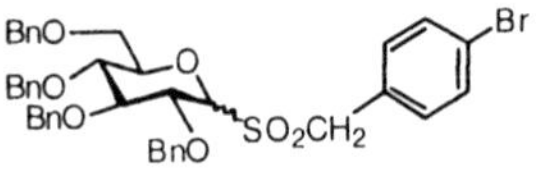 BnO, BnO, BnO, O, BnO, SO_2CH_2, Br	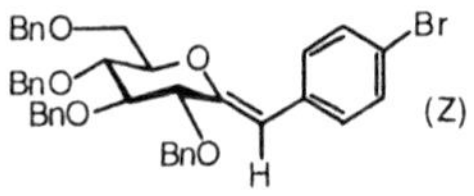 BnO, BnO, BnO, O, BnO, H, Br (Z)	84% Z/E=94/6
5	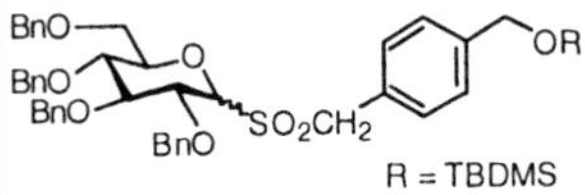 BnO, BnO, BnO, O, BnO, SO_2CH_2, OR R = TBDMS	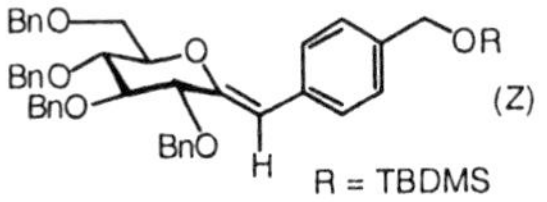 BnO, BnO, BnO, O, BnO, H, OR (Z) R = TBDMS	88% Z/E=80/20
6	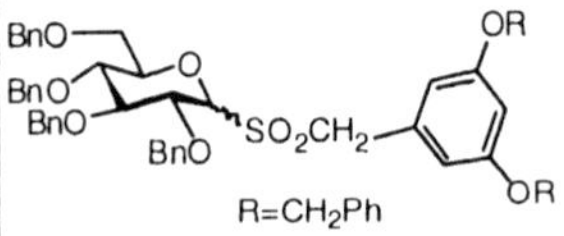 BnO, BnO, BnO, O, BnO, SO_2CH_2, OR, OR R=CH_2Ph	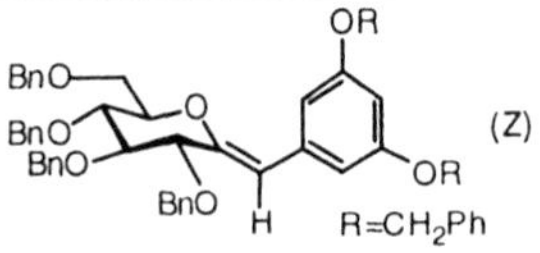 BnO, BnO, BnO, O, BnO, H, OR, OR (Z) R=CH_2Ph	82 Z/E = 2.2:1
7	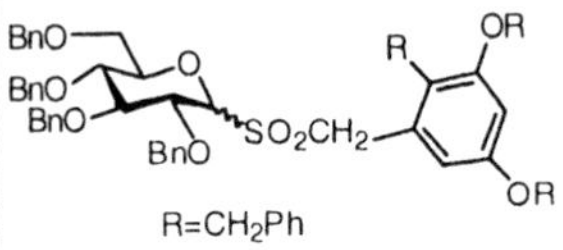 BnO, BnO, BnO, O, BnO, SO_2CH_2, R, OR, OR R=CH_2Ph	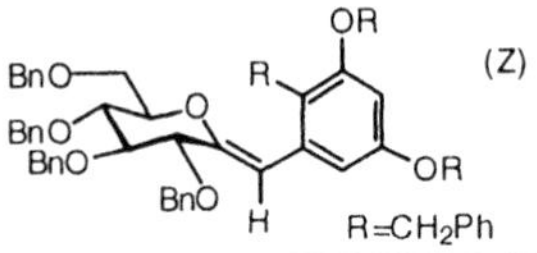 BnO, BnO, BnO, O, BnO, H, R, OR, OR (Z) R=CH_2Ph	87 Z/E = 2.9:1

Table 2 Examples of Ramberg-Backlund reactions of benzylic glycosyl sulfones

	sulfone	Ramberg-Backlund product	yields, E/Z ratio, notes
8	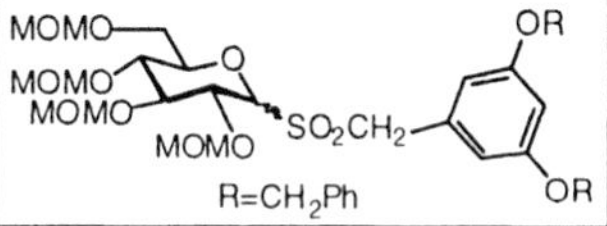 R=CH_2Ph	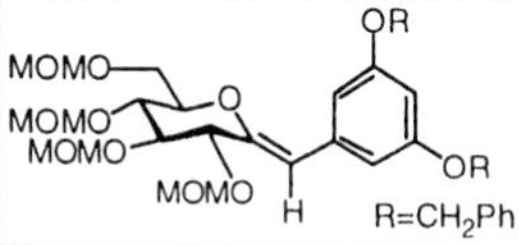 (Z) R=CH_2Ph	87% Z/E=2.5/1
9	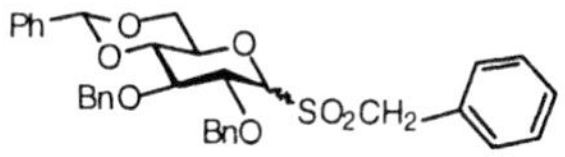	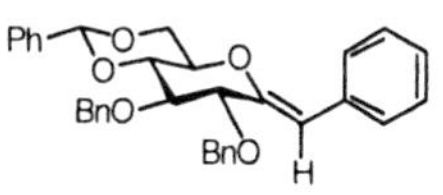 (Z)	88% Z/E=6/1
10	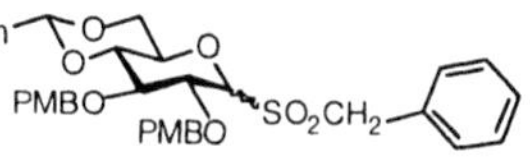	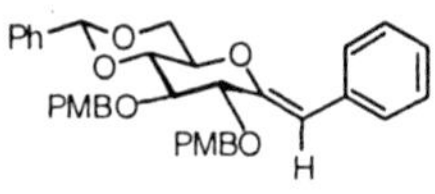 (Z)	88% Z/E=6/1
11	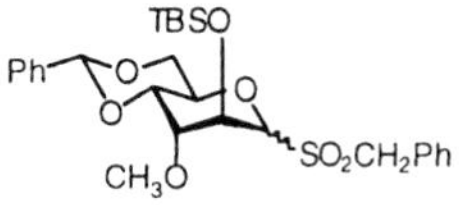	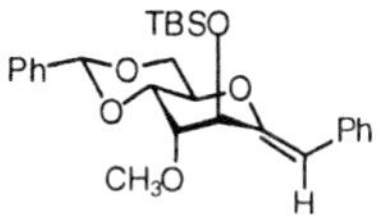 (Z)	70% Z only
12	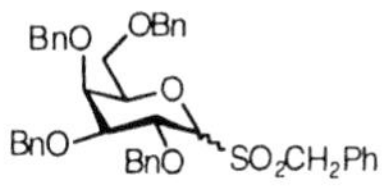	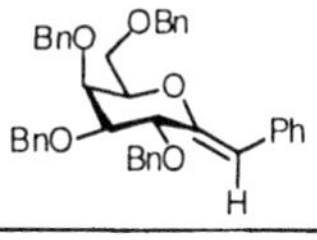 (Z)	58% Z/E=56/44 with $C_3F_6Br_2$ 64%, Z/E n.d.
13	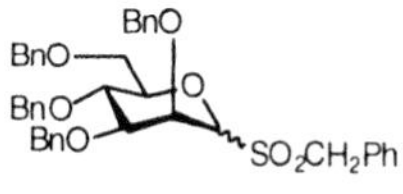	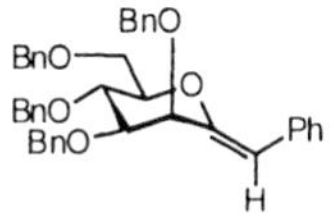 (Z)	62% Z/E=95/5
14	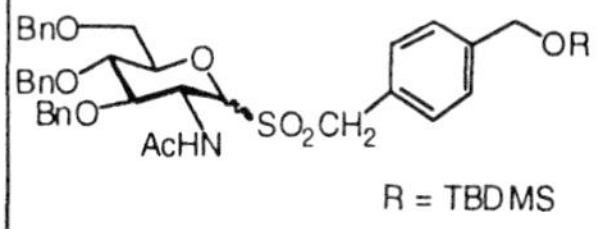 R = TBDMS	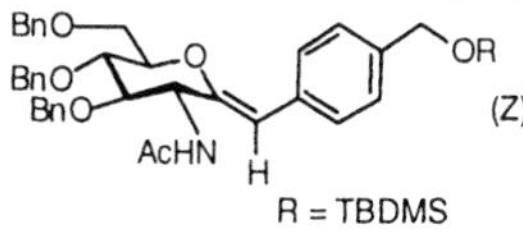 (Z) R = TBDMS	65% Z/E n.d.

Table 3 Examples of Ramberg-Backlund reactions of glycosyl alkylsulfones

	sulfone	base/ "Br"	Ramberg-Backlund products	
15	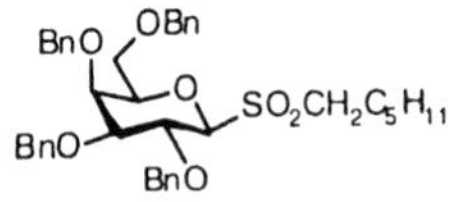 BnO, OBn, O, $SO_2CH_2C_5H_{11}$, BnO, BnO	KOH/Al_2O_3 CF_2Br_2 0°	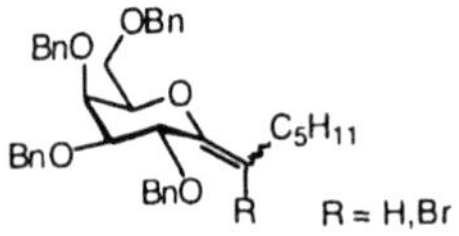 BnO, OBn, O, C_5H_{11}, BnO, BnO, R; R = H,Br	(a) CH_2Cl_2 R=H Z11%E13% R=Br Z6%E35% **b) THF** R=H E7% R=Br Z67% E8% c)DME R=H E5% R=Br Z80%E11%
16	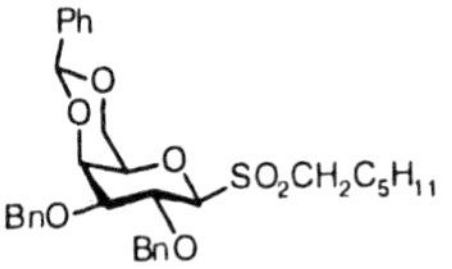 Ph, O, O, O, $SO_2CH_2C_5H_{11}$, BnO, BnO	KOH/Al_2O_3 CF_2Br_2 25° KOH/Al_2O_3 $C_2F_4Br_2$ 45°	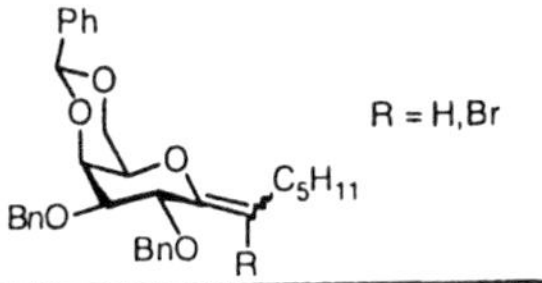 Ph, O, O, O, C_5H_{11}, BnO, BnO, R; R = H,Br	**a) CH_2Cl_2 R=Br 88% Z/E nd** b) *t*-BuOH R=H 83% Z/E=40/60
17	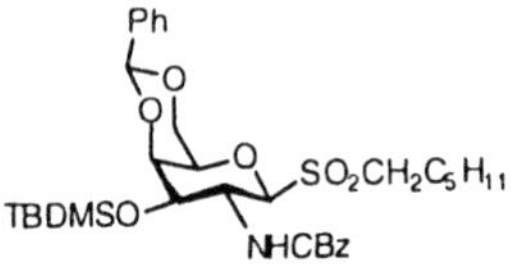 Ph, O, O, O, $SO_2CH_2C_5H_{11}$, TBDMSO, NHCBz	KOH/Al_2O_3 CF_2Br_2 5° KOH powder CCl_4 60°	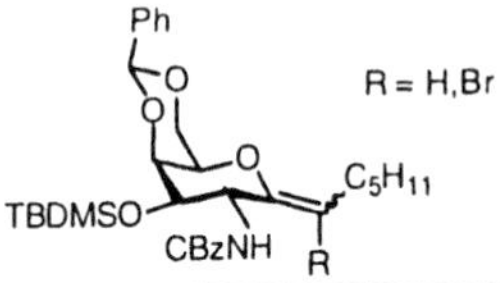 Ph, O, O, O, C_5H_{11}, TBDMSO, CBzNH, R; R = H,Br	a) R=Br 35% Z/E nd b) *t*-BuOH R=H 15% Z/E nd
18	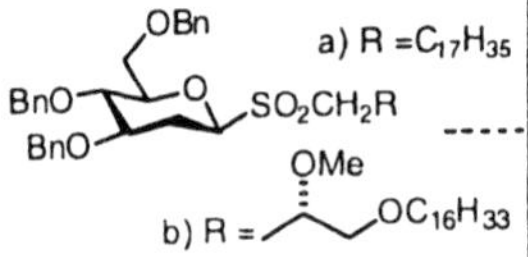 OBn, BnO, BnO, O, SO_2CH_2R; a) R = $C_{17}H_{35}$; b) R = OMe, $OC_{16}H_{33}$	KOH/Al_2O_3 $C_2F_4Br_2$ 45° KOH/Al_2O_3 $C_2F_4Br_2$ 45°	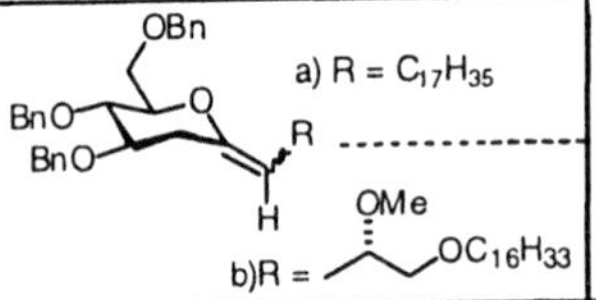 OBn, BnO, BnO, O, R, H; a) R = $C_{17}H_{35}$; b)R = OMe, $OC_{16}H_{33}$	a) *t*-BuOH 85% Z/E=1/1 b) *t*-BuOH 51% Z/E=1/1 9% endo
19	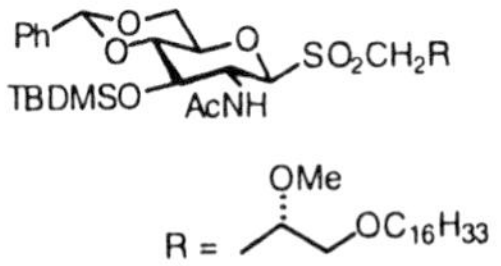 Ph, O, O, O, SO_2CH_2R, TBDMSO, AcNH; R = OMe, $OC_{16}H_{33}$	KOH/Al_2O_3 $C_2F_4Br_2$ 45°	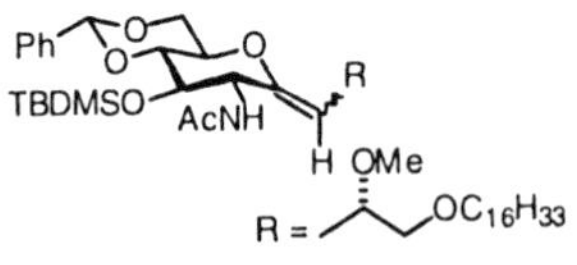 Ph, O, O, O, R, TBDMSO, AcNH, H; R = OMe, $OC_{16}H_{33}$	*t*-BuOH 70% Z only
20	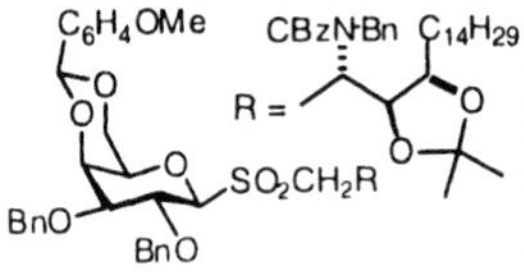 C_6H_4OMe, O, O, O, SO_2CH_2R, BnO, BnO; R = CBzN-Bn, $C_{14}H_{29}$, O, O	KOH/Al_2O_3 $C_2F_4Br_2$ 45°	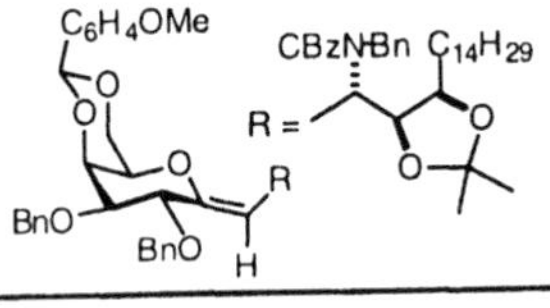 C_6H_4OMe, O, O, O, R, BnO, BnO, H; R = CBzN-Bn, $C_{14}H_{29}$, O, O	*t*-BuOH 76% Z/E nd

Scheme 2 Rationale for differing reactivities of α and β Sulfones

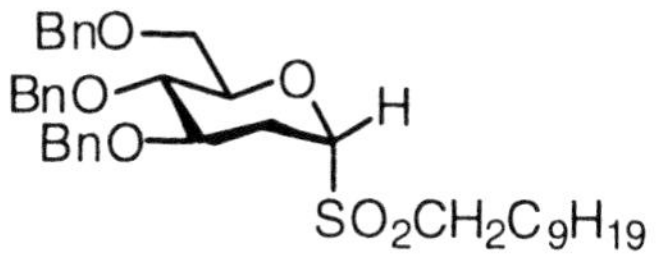

BnO BnO BnO O H $SO_2CH_2C_9H_{19}$

12a

BnO BnO BnO O H H S O O RCH_2

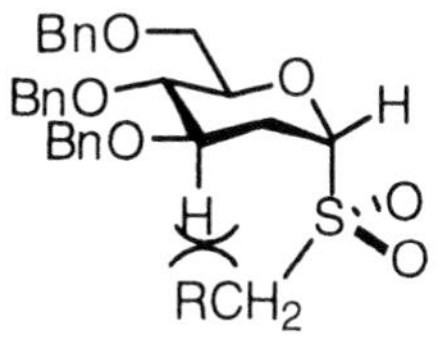

a

best conformation
to support stereoelectronic
enhancement of H-1 acidity
in α isomer,but there is severe
steric hindrance

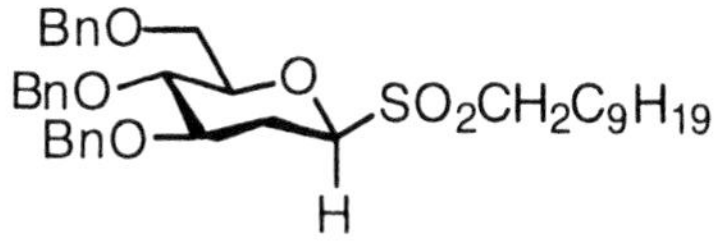

BnO BnO BnO O $SO_2CH_2C_9H_{19}$ H

12b

BnO BnO BnO O CH_2R S O O H H

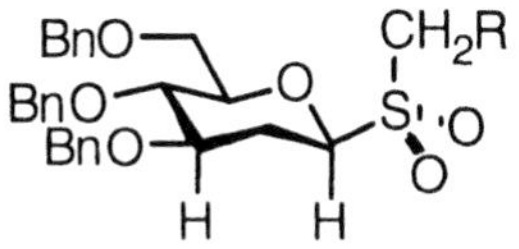

b

best conformation
to support stereoelectronic
enhancement of H-1 acidity in
β isomer, so it does react

A second example of disappointing reactivity is that of the disaccharide sulfone **13** (Fig. 3). The desired R-B product **14** is obtained via the KOH/CF_2Br_2 method, but in less than 5% yield. This should be compared to the report of a similar disaccharide synthesis by Taylor where a better choice of conditions (KOH/CCl_4/60°) afforded the product in 48% yield.(*19*)

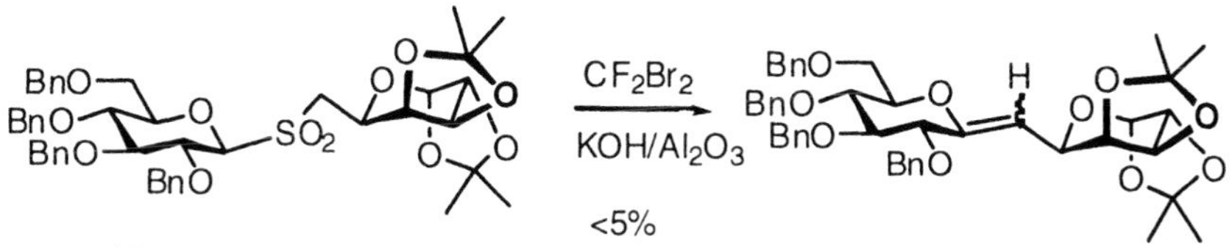

BnO BnO BnO BnO O SO_2 O O O O O

13

CF_2Br_2 → KOH/Al_2O_3 <5%

BnO BnO BnO BnO O H O O O O O

14

Fig. 3

In two cases, elimination of the substituent at C-2 of galactose derivatives was observed. With the tetrabenzyl galactose derivative **15**, the product glycal **16** was formed to the exclusion of the desired R-B product when the $C_2F_4Br_2$ reagent was used (Fig. 4). This may be contrasted to the very different results reported in entry 15, Table 3 for the same sulfone and the lower temperature conditions used with the CBr_2F_2 reagent. Also, in entry 16, where the galactose is rigidified by the benzylidene ring, the R-B using $C_2F_4Br_2$ is quite successful. This cluster of experiments highlights how closely balanced are the energetics for the competing pathways of base-catalyzed bromination, β-elimination of the C-2 substituent and cycloelimination to form the R-B product.

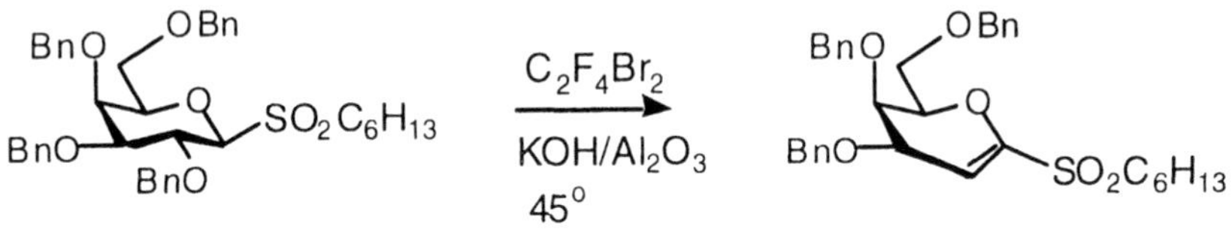

BnO OBn BnO BnO $SO_2C_6H_{13}$ **15** → $C_2F_4Br_2$, KOH/Al_2O_3, 45° → BnO OBn BnO $SO_2C_6H_{13}$ **16**

Fig. 4

Also, the rigidification of the galactose ring was not sufficient to block the elimination of the 2-azido substituent of **17** even using the low-temperature reagent (Fig. 5). Another failure occurred with the isoxazole sulfone **19**. Here, the heterocycle did not survive base treatment. The isoxazole sulfide precursor was also unstable to base (Fig. 6).

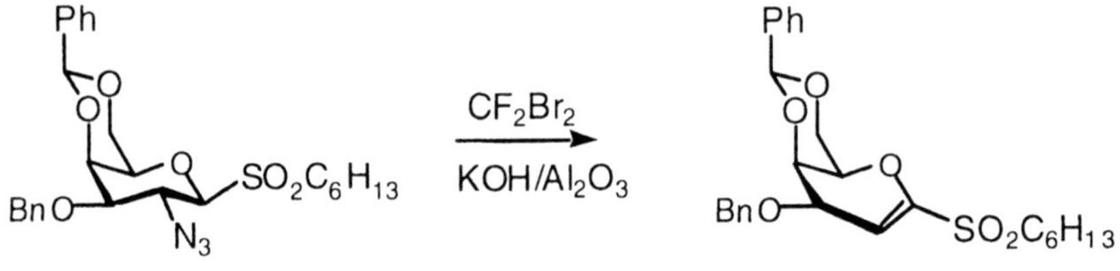

Ph BnO N_3 $SO_2C_6H_{13}$ **17** → CF_2Br_2, KOH/Al_2O_3 → Ph BnO $SO_2C_6H_{13}$ **18**

Fig. 5

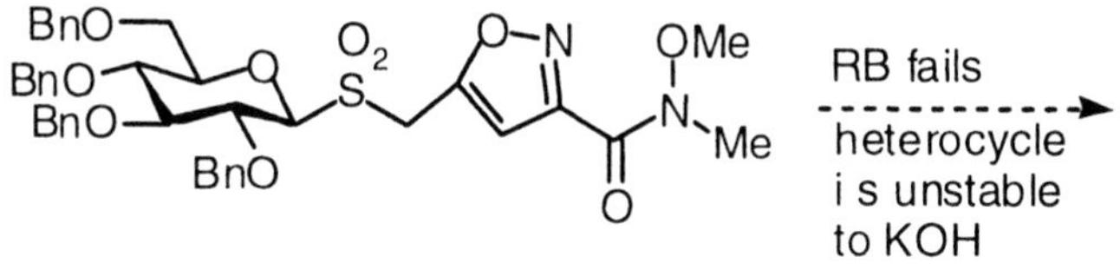

RB fails
heterocycle is unstable to KOH

19

Fig. 6

Further processing of the exo-glycal products is required to obtain target C-glycosides.(*25-27*) Most commonly, hydrogenation with Pd/C catalysts was used to produce β-C-glycosides in good yield (i.e. hydrogen delivery from the axial face of C-1) exemplified by the conversion of **20** to **21** (Fig. 7). It was usually the case that the glycal double bond was reduced much more rapidly than the hydrogenolysis of benzyl groups. In the 2-deoxy series, entries 1,2 and 18, there was also some Pd-catalyzed double bond migration to the endocyclic position and this alkene was hydrogenated at a slower rate.

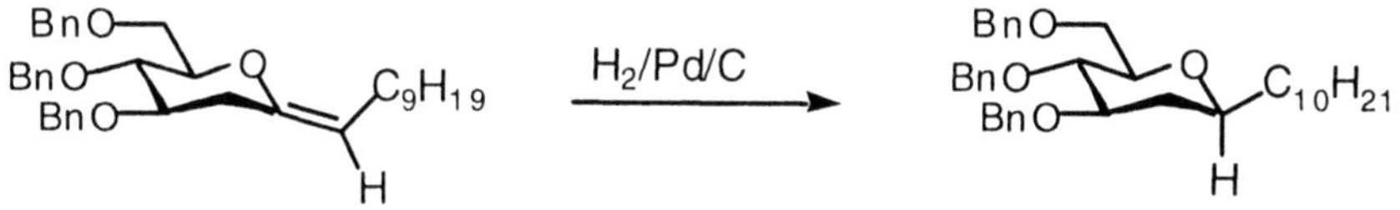

$H_2/Pd/C$

20 **21**

Fig. 7

An alternate method, hydride transfer from Si-H reagents to a carbonium-like C-1 was also developed as a reduction process (Scheme 3). This method also produced β-glycosides **26** except in those instances where the Si-H function was attached to the 4-OH of galactose species. In those cases (derived from entries 16b and 20), the hydride was delivered from the top face and the final product was an α-C-glycoside **25**. The preferred sequence of events in the hydride tranfer method was to first react the exo glycal with MeOH and a trace of acid generated by addition of a drop of TMSCl. This led to stable α-OMe glycosides **23**. Then, at the appropriate juncture, the OMe was activated by BF_3 which led to the formation of the C-1 carbonium ion **24** which would then be instantly trapped by the hydride donor. This process was key to our successful synthesis of an α-C-galactosylceramide, a potent immunostimulant.(*22,27*)

Scheme 3 C-1 Oxacarbenium ion reduction by Si-H species

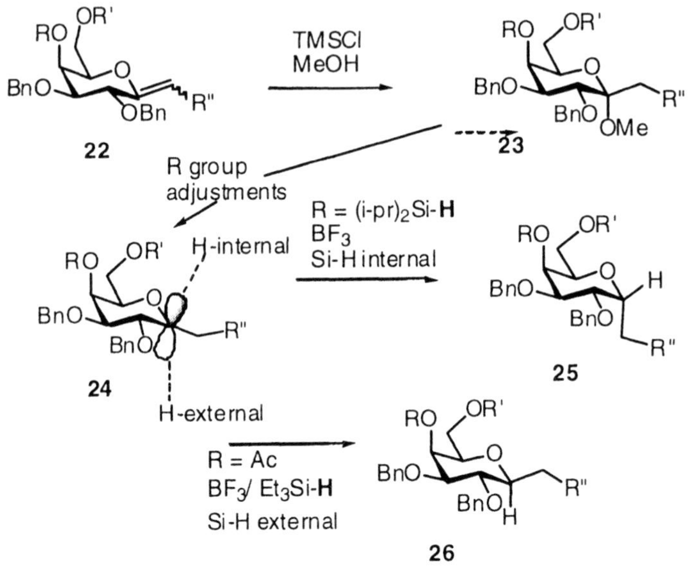

An important conversion of exo-glycals for our work on altromycin was hydroboration (Scheme 4). We observed that the exo glucal **27** of entry 10 led to a 3/1 ratio of α(**28**) and β C-glycosides. By contrast, similar treatment of the altrose-derived glycal **29** of entry 11 afforded a 1/10 ratio. Since our program required an α-C-glycoside, the product **28** was carried further to finally produce an altromycin model compound.(*21*)

Scheme 4 Hydroboration of exo glycals

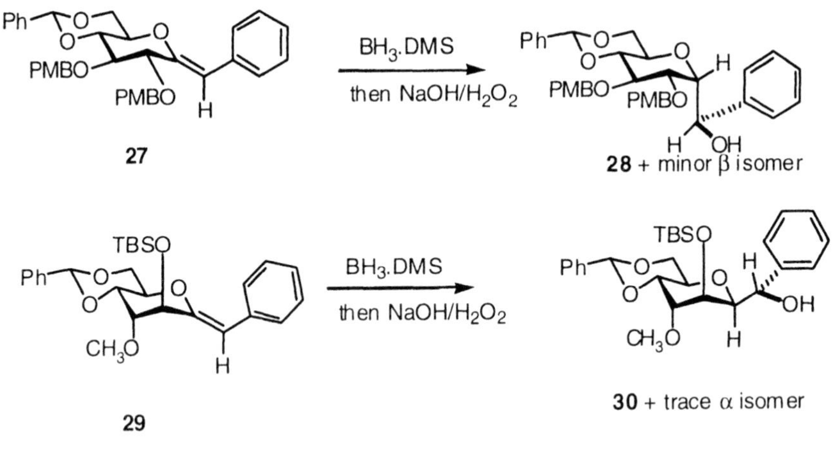

In conclusion, the Ramberg- Bäcklund reaction of glycosyl sulfones is a useful and general route to a variety exo glycals. Simple conversions of the exo-glycals make available a variety of C-glycosides.

Experimental Section

Preparation of KOH/Alumina reagent A batch of 30 g of KOH pellets was dissolved in 300 mL of MeOH and slurried with 90 g of neutral alumina (E. Merck, grade 1, 60 mesh), then the solvent was removed *in vacuo* until the alumina was again free flowing. When kept in a stoppered container, this material sustained a shelf life of several months.

Preparation of 2, 6-anhydro-1-deoxy-1-phenyl-3, 4, 5, 7-tetrakis-O-(phenylmethyl)-D-manno-hept-1-enitol A solution of α-tetrabenzylmannosyl benzylsulfone (entry 13, Table 2) (1.3 g, 1.92 mmol) in *tert*-butanol (10 mL) and dichloromethane (10 mL) was cooled to 5 °C and mixed with alumina supported potassium hydroxide (8 g, ~48 mmol). Dibromodifluoromethane (2.5 mL, ~28 mmol) was added dropwise over five seconds and the reaction mixture was stirred at 5 °C. After 5 min, TLC analysis (30% EtOAc:hexane) showed mostly alkene product with some intermediate bromosulfone and some presumably unreacted α–sulfone. Stirring of the reaction mixture at 5 °C was continued for an additional 1 h and TLC analysis showed only a trace of unreacted intermediates. The reaction mixture was diluted with dichloromethane and filtered through a pad of Celite. The solids were washed well with dichloromethane and the filtrate was concentrated under reduced pressure to dryness to give 1.6 g of an oil. The oil was slurried in a small amount of toluene and flash chromatographed on silica gel 60 (15 g) packed in hexane. Elution with 5% ethyl acetate:hexane gave alkene (entry 13, Table 2) as a colorless oil (730 mg, 61.5% yield) as a 95:5 *Z*:*E* alkene mixture: ^{1}H NMR (300 MHz, $CDCl_3$) δ 3.74 (dd, 1H, J = 3.2, 9.2 Hz), 3.78-3.83 (ddd, 1H), 3.86-3.96 (m, 2H), 4.16 (d, 1H, J = 3 Hz, H-3), 4.24 (t, 1H, J = 9 Hz), 4.47-4.74 (m, 6H, Ar$\underline{CH_2}$), 4.81 (1/2 AB_q, 1H, J = 12.9 Hz, Ar$\underline{CH_2}$), 5.00 (1/2 AB_q, 1H, J = 11.1 Hz, Ar$\underline{CH_2}$), 5.58 (s, 0.95H, H-1_Z), 6.58 (s, 0.05H, H-1_E), 7.18-7.46 (m, 23H, Ar$\underline{H}$), 7.72 (d, 2H, J = 7.3 Hz, Ar$\underline{H}$); ^{13}C NMR (75 MHz, $CDCl_3$) δ 69.02 (t), 69.68 (t), 71.28 (t), 73.16 (t), 74.13 (d), 74.89 (d), 74.98 (t), 79.63 (d), 81.70 (d), 114.86 (d), 127.06 (d), 127.47 (d), 127.62 (d), 127.72 (d), 127.87 (d), 128.12 (d), 128.32 (d), 128.90 (d), 129.00 (d), 134.46 (s), 138.06 (s), 138.12 (s), 138.34 (s), 138.40 (s), 148.49 (s);(+) ESMS *m/e* 635 $[M + Na]^+$.

(3S)-3-*O*-methyl-4-*O*-hexadecyl-2′-acetamido-4′,6′-*O*-benzylidene-3′-*O*-(tert-butyldimethylsilyl)-2′-deoxy-D-glucopyranosylidene butane To a solution of sulfone (entry 19, Table 3) (0.12 g, 0.15 mmol) in *t*-BuOH (1.5 mL) and CF_2BrCF_2Br (2 mL) was added 0.3 g (25% by weight) of KOH/Al_2O_3 (prepared one day earlier). The mixture was heated at 47 °C overnight. The solution was filtered through a pad of Celite which was washed with CH_2Cl_2. The residue was purified by column chromatography on silica gel (eluting with 40% EtOAc-PE) to afford 0.056 g (70%, Z isomer only) of (entry 19, Table 3) as a colorless oil. ^{1}H NMR (500 MHz, CDCl3): δ 7.48-7.45 (m, 2H), 7.37-7.34 (m, 3H), 5.54 (s, 1H), 5.38 (d, J = 9.3 Hz, 1H, NH), 4.88 (t, J = 6.8 Hz, 1H, vinyl H), 4.62 (t, J = 8.8 Hz, 1H, H-2), 4.38 (dd, J = 5.1, 10.5 Hz, 1H, H-6), 3.80 (t, J = 10.3 Hz, 1H, H-6), 3.63 (m, 2H), 3.43-3.34 (m, 9H), 2.45 (m, 1H), 2.24 (m, 1H), 2.05 (s, 3H, NAc), 1.57 (m, 2H), 1.25 (s, 26H), 0.87 (t, J = 6.6 Hz, 3H, CH_3), 0.82 (s, 9H), 0.03 (s, 3H, CH_3), -0.04 (s, 3H, CH_3); ^{13}C NMR ($CDCl_3$, 75 MHz): δ 169.46, 150.38, 137.15, 129.19, 128.25, 126.41, 105.65, 102.09, 82.26, 79.83, 74.59, 72.63, 71.96, 70.64, 68.95, 57.48, 54.59, 32.19,29.96, 29.79, 29.63, 26.43, 26.33, 25.91, 23.84, 22.96, 18.37, 14.39, -3.66, -4.54; MS: *m/z* 754 (M^++Na^+), (calcd. $C_{42}H_{73}O_7NSi$, 731).

References

1. Ramberg, L.; Bäcklund, B. *Ark. Kemi Mineral. Geol.* **1940**, *13A*, no. 27, 1; *Chem. Abstr.* **1940**, *34*, 4725.
2. For reviews on the Ramberg-Bäcklund reaction, see (*2-8*): Bordwell, F. G. In *Organosulfur Chemistry*; Janssen, M. J., Ed.; Wiley: New York, **1967**, Chapter 16, p 271.
3. Paquette, L. A. *Acc. Chem. Res.* **1968**, *1*, 209.
4. Paquette, L. A. *Org. React.* **1977**, *25*, 1.
5. Clough, J. M. In *Comprehensive Organic Synthesis*; Trost, B. M., Fleming, I., Pattenden, G., Eds.; Pergamon Press: Oxford, **1991**; Vol. 3, Chapter 3.8.
6. Taylor, R. J. K. *Chem. Commun.* **1999**, 217.
7. Paquette, L. A. *Synlett* **2001**, 1.
8. Casy, G.; Taylor, R.J.K. *Organic React.* **2003**, *62*, 357.
9. Acton, E.M.; Ryan, K.J.; Tracy, M.; Arora, S.K. *Tetrahedron Lett.* **1986**, *27*, 4245-4248.
10. Welch, S.C.; Levine, J.A.; Arimilli, M.N. *Synth. Comm.* **1993**, *23*, 131-134.
11. Hart, D.J.; Merriman, G.H.; Young, D.G.J. *Tetrahedron*, **1996**, *52*, 14437-14458.
12. For applications of the Ramberg-Bäcklund reaction to the synthesis of C-glycosides reported by our group and by the Taylor group, see refs (*12-22*): Belica, P. S.; Franck, R. W. *Tetrahedron Lett.* **1998**, *39*, 8225.
13. Yang, G.; Franck, R. W.; Byun, H.-S.; Bittman, R.; Samadder, P.; Arthur, G. Org. Lett. **1999**, 1, 2149.
14. Yang, G.; Franck, R. W.; Bittman, R.; Samadder, P.; Arthur, G. *Org. Lett.* **2001**, *3*, 197.

15. Pasetto, P.; Chen, X.; Drain, C. M.; Franck, R. W. *Chem. Commun.* **2001**, 81.
16. Pasetto, P.; Franck, R.W. *J. Org. Chem.* **2003**, *68*, 8042.
17. Griffin, F. K.; Murphy, P. V.; Paterson, D. E.; Taylor, R. J. K. *Tetrahedron Lett.* **1998**, *39*, 8179.
18. Alcaraz, M.-L.; Griffin, F. K.; Paterson, D. E.; Taylor R. J. K. *Tetrahedron Lett.* **1998**, *39*, 8183.
19. Paterson, D. E.; Griffin, F. K.; Alcaraz, M.-L.; Taylor, R. J. K. *Eur. J. Org. Chem.* **2002**, 1323.
20. Ohnishi, Y.; Ichikawa, Y. *Bioorg. Med. Chem. Lett.* **2002**, *12*, 997.
21. Griffin, F. K.; Paterson, D. E.; Murphy, P. V.; Taylor, R. J. K. *Eur. J. Org. Chem.* **2002**, 1305.
22. Yang, G.; Schmieg, J; Tsuji, M. Franck, R.W.; *Angew. Chemie Int. Ed.* **2004**;*43*, 0000.
23. Chan, T.-L.; Fong, S.; Li, Y.; Man, T.-O.; Poon, C-D. *Chem. Commun.* **1994**, 1771-1772.
24. Simpkins, N. S. Sulphones in organic synthesis Oxford [England] ; New York : Pergamon, **1993**.
25. For recent references to applications of exo glycals, see (*25-27*): Yang, W.-B.; Yang, Y.-Y.; Gu, Y.-F.; Wang, S.-H.; Chang, C.-C.; Lin, C.-H.; *J. Org. Chem.*; **2002**; *67*; 3773-3782.
26. Gascón-López, M.; Motevalli, M.; Paloumbis, G.; Bladon, P.; Wyatt, P.B. *Tetrahedron*, **2003**, *59*, 9349-9360.
27. Taillefumier, C.; Chapleur, Y. *Chem. Reviews* **2004**, *104*, 263-292.
28. Schmieg, J.; Yang, G.; Franck, R.W.; Tsuji, M. *J. Exp. Med.* **2003**;*198* 1631-1641.

Chapter 7

Conformationally Restrained *C*-Glycoside Probes for Sialyl Lewis X-Selectin Recognition

R. W. Denton and D. R. Mootoo

Department of Chemistry, Hunter College and Graduate Center, City University of New York, 695 Park Avenue, New York, NY 10021

The interaction of tetrasaccharide sialyl Lewis X (sLe^x) **1** and the selectins is an early event in the inflammation response. Small molecule mimetics of sLe^x have received attention as probes of this recognition process, and conformationally restrained templates are appealing because of their potential to provide precise structural information. Herein, we report the synthesis of two conformationally restricted C-glycoside scaffolds related to the 1,1-linked Gal-Man disaccharide **2**, a known mimetic of sLe^x. The synthetic plan centers on formation of the glycone segment via an enol ether-oxocarbenium ion cyclization and starts from 1-thio-2,3-O-isopropylidene (TIA) acetal **16** and carboxylic acids **14** and **15**. The method constitutes a general one for β-C-galactosides with complex aglycone segments.

Introduction

The inflammation response to tissue injury involves infiltration of damaged areas by leukocytes from the blood stream *(1-4)*. Unregulated extravasation of leukocytes can result in inflammatory disorders such as reperfusion injuries, stroke, psoriasis, rheumatoid arthritis and respiratory diseases. An early step in the cascade of events leading to influx of leukocytes is the recognition of the tetrasaccharide sLe^x **1** (found on the terminus of leukocyte surface glycoproteins), by E, P and L selectins that are expressed by endothelial cells lining blood vessels (Fig. 1). Since E and P selectins on endothelial cells are activated or upregulated in response to injury, modulation of selectin-sLe^x binding could provide an effective strategy for treating inflammation disorders *(5, 6)*. In this vein, small molecule mimetics of sLe^x that can act as selectin antagonists, have received considerable attention *(7-13)*. An attractive lead compound from the Wong group is the 1,1-linked Gal-Man disaccharide **2** *(14, 15)*. Although much simpler than sLe^x, compound **2** was reported to be more active against E- and P- selectin.

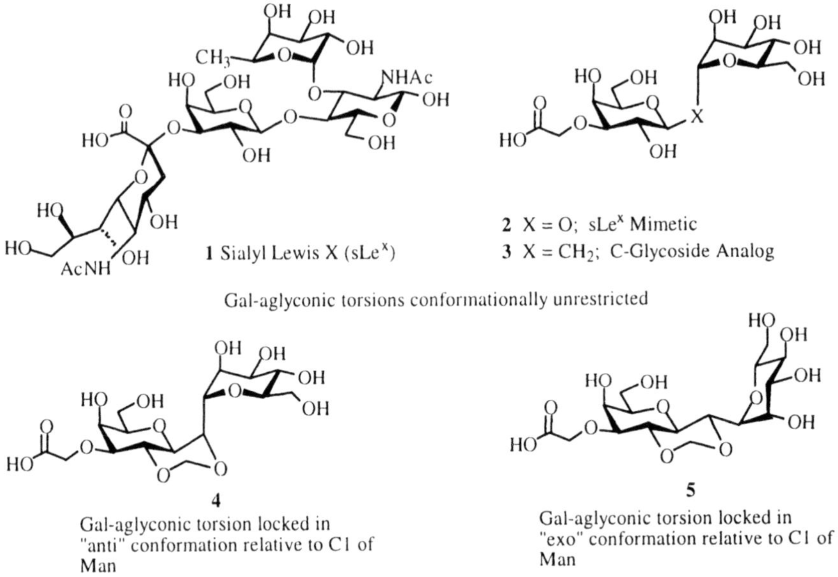

1 Sialyl Lewis X (sLe^x)

2 X = O; sLe^x Mimetic
3 X = CH_2; C-Glycoside Analog

Gal-aglyconic torsions conformationally unrestricted

4
Gal-aglyconic torsion locked in "anti" conformation relative to C1 of Man

5
Gal-aglyconic torsion locked in "exo" conformation relative to C1 of Man

Fig. 1: sLe^x and C-glycoside analogues of sLe^x Mimetic **2**

Inspection of the popular model for sLe^x-selectin binding, suggests that the Gal and Fuc recognition sites for sLe^x are homogeneous with the binding regions for the Gal and Man residues in **2**, respectively *(16-18)*. Accordingly, analogues of **2** in which the Man residue occupies restricted regions of conformational space, could be used as probes for structural information on the contacts in the Fuc binding region of the selectins. In so far as the binding of the Fuc residue is believed to account for the major part of the overall sLe^x-selectin binding energy *(19-23)*, this information would be valuable in the design of more potent selectin antagonists. Against this backdrop, we undertook a program aimed at the synthesis of C-glycosides **3-5**, and the investigation of their conformational and selectin binding properties. Our initial efforts led to the synthesis of the less rigid C-glycoside **3** *(24)*, and the finding that this C-disaccharide scaffold samples five conformational families with respect to the intersaccharide linkage *(25)*. By comparison, the O-glycoside **2** exists in greater than 95% in a single conformation ("exo-exo"). Analogues **4** and **5** are interesting because they have restrained conformations with respect to the Gal glyconic bond. Consequently, **4** and **5** present the Man residue in a more well defined region of conformational space compared to **3**, and could provide more precise structure activity information *(26, 27)*. Herein, we describe our synthetic studies on C-disaccharide precursors to **4** and **5**.

Results and Discussion

Retrosynthesis

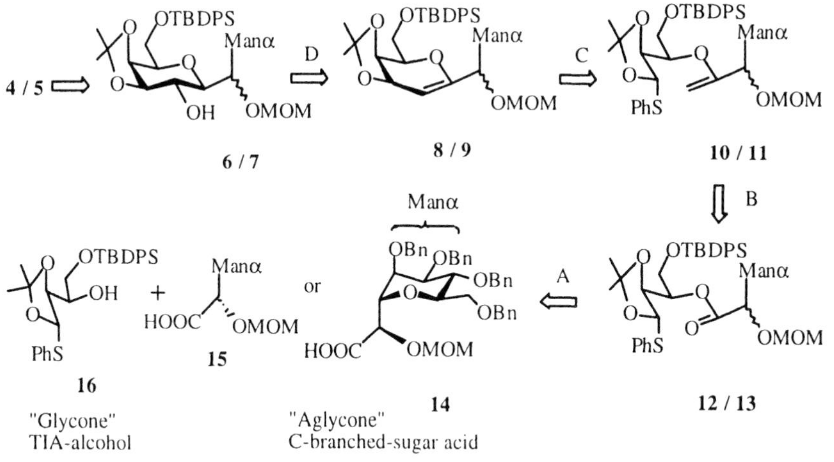

A: Esterification; B: Tebbe methylenation; C: Oxocarbenium ion cyclization; D: Hydroboration

Scheme 1: β-C-galacto-disaccharides from TIA's - Retrosynthesis of **4/5**

The synthesis of **4** and **5** provided an opportunity to evaluate the scope of a new C-glycosidation methodology that was used for **3** *(28-34)*. Accordingly, esterification (step A) of the "glycone" component, 1-thio-1,2-O-isopropylidene acetal (TIA) **16** and one or the other "aglycone" segments, C-branched saccharide acids **14** or **15**, furnishes ester **12** or **13**, respectively (Scheme 1). Tebbe methylenation (step B) of the latter provides enol ethers **10** or **11**. Thiol activation in **10** and **11** leads to the corresponding oxocarbenium ions, and ensuing oxocarbenium ion cyclization (step C) involving the pendant enol ethers, gives the corresponding C1 substituted glycals **8** or **9**. Stereoselective hydroboration (step D) of **8** or **9** affords β-C-galacto-disaccharides **6** or **7**, which could be transformed to the eventual target structures **4** and **5** through standard alcohol group manipulations. Attributes of this strategy are the high level of convergency, and the compatibility with structures containing complex aglycone segments. However, although the four-step sequence (i.e A–D) was previously successful for a variety of methylene linked C-disaccharides such as **3**, the protocol had not been tested on structures with an alkoxy substituent on the intersaccharide carbon (e.g. **6** and **7**). We were concerned that steric and electronic changes brought about by this modification could have a deleterious effect.

Synthesis of "glycone" and "aglycone" segments

The details on the preparation of the "glycone" synthon **16** have been described in the earlier report on **3** *(24)*. The key reactions in this synthesis is the Suarez radical fragmentation of the 1,2-O-isopropylidene furanose **17** to give the 1-O-acetyl-1,2-isopropylidene **18** *(35)*, and acetal exchange of **18** to the phenylthio monothioacetal **16** (Scheme 2). Overall, **16** is easily obtained on multi-gram scale (ca. 20 g), in 60% yield, over five steps from commercially available D-lyxose.

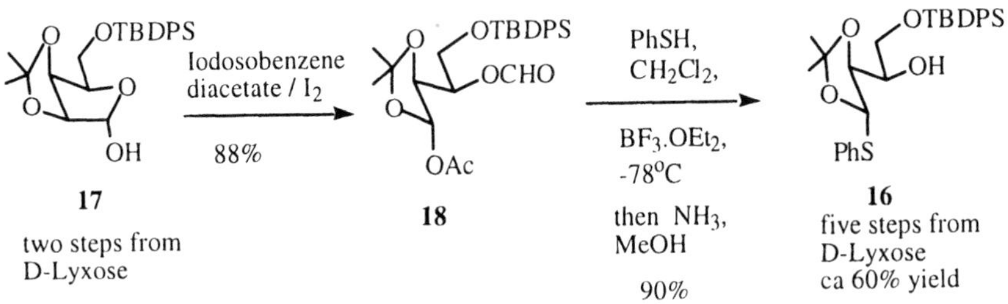

OTBDPS, O, O, O, OH — **17** — two steps from D-Lyxose

Iodosobenzene diacetate / I_2 — 88%

OTBDPS, OCHO, O, O, OAc — **18**

PhSH, CH_2Cl_2, $BF_3.OEt_2$, -78°C then NH_3, MeOH — 90%

OTBDPS, OH, O, O, PhS — **16** — five steps from D-Lyxose ca 60% yield

Scheme 2: Synthesis of the "glycone" synthon

The "aglycone" components **14** and **15** were prepared via a straightforward sequence of reactions from the known C-formylmannoside **19**, available in four steps from methyl α-D-mannopyranoside *(14)*. Treatment of **19** with vinylmagnesium bromide provided a mixture of alcohols that was separated as the methoxymethyl ether derivatives **20** and **21** (63% from **19**, 3:2 respectively, Scheme 3). The configuration at the allylic position in **20** and **21** was assigned in the conformationally restrained derivatives **31** and **32** (*vide infra*). Ozonolysis of **20** and **21**, followed by sodium chlorate oxidation of the resulting aldehydes led to **14** and **15**. Two additional details regarding the synthesis of **14** ad **15**, are noteworthy. First, due to concern about the configurational stability of the C1 aldehyde, it was necessary to confirm that the axial orientation of the C1 branch was conserved in **20** and **21**. Therefore, **20** and **21** were individually transformed to their tetraacetylated-dihydro derivatives, and the stereochemical integrity of the latter verified by NOESY experiments. Second, due to the very similar chromatographic mobility of **20** and **21**, it was generally more practical to first convert the mixture of alcohols from the Grignard reaction to their methoxyacetate derivatives. This allowed for a simpler chromatographic resolution, following which pure samples of **20** and **21** could be obtained through hydrolysis of the separated esters and methoxymethylation of the resulting alcohols.

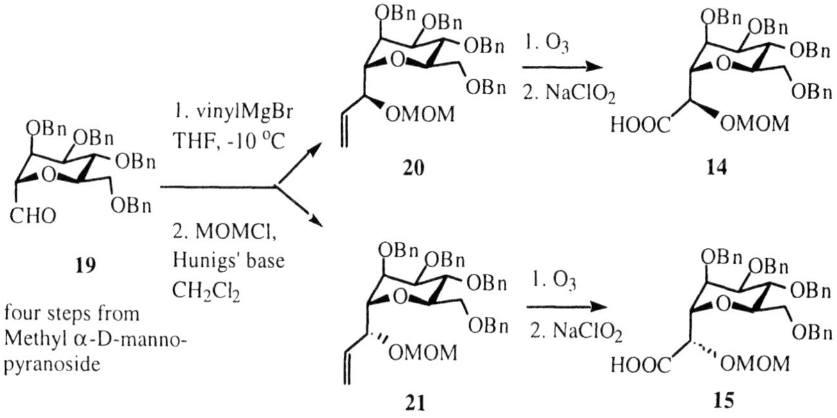

OBn OBn OBn OBn CHO **19** four steps from Methyl α-D-mannopyranoside

1. vinylMgBr THF, -10 °C 2. MOMCl, Hunigs' base CH_2Cl_2

OBn OBn OBn OBn OMOM **20** — 1. O_3 2. $NaClO_2$ → OBn OBn OBn OBn HOOC OMOM **14**

OBn OBn OBn OBn OMOM **21** — 1. O_3 2. $NaClO_2$ → OBn OBn OBn OBn HOOC OMOM **15**

Scheme 3: Synthesis of the "aglycone" components

C-glycoside synthesis

The aglycone segments **14** and **15** were next individually partnered with "glycone" **16** for the four step C-glycosidation sequence (Scheme 4). DCC mediated esterification *(36)* of **16** with **14** and **15** followed by treatment of the resulting esters with the Tebbe reagent *(37)* gave enol ethers **23** and **25** respectively, in 61 and 65% yields from **16**. The key cyclization reactions on **23** and **25** to C1 substituted glycals **8** (82%) and **9** (92%) respectively, were promoted by methyl triflate in the presence of 2,6-di-*tert*-butyl-4-methylpyridine. Hydroboration of the glycals provided β-C-galactosides **6** (88%) and **7** (86%), each as a single diastereomer. The stereochemistry of these products was confirmed by ^{1}H NMR analysis of later derivatives (*vide infra*). These results suggest that the four step C-glycosidation protocol is compatible with complex α-alkoxy acid derivatives, and opens the way to preparation of a variety of cyclic ether motifs.

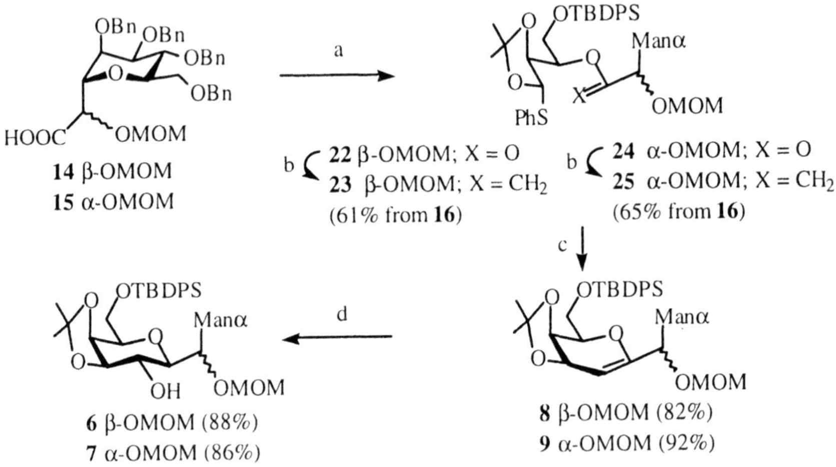

14 β-OMOM
15 α-OMOM

22 β-OMOM; X = O
23 β-OMOM; X = CH_2
(61% from **16**)

24 α-OMOM; X = O
25 α-OMOM; X = CH_2
(65% from **16**)

8 β-OMOM (82%)
9 α-OMOM (92%)

6 β-OMOM (88%)
7 α-OMOM (86%)

(a) **16**, DCC, DMAP, PhH; (b) Tebbe; (c) MeOTf, DTBMP, CH_2Cl_2; (d) BH_3.DMS then Na_2O_2

Scheme 4: Synthesis of β-C-galacto-disaccharides **6/7**

Introduction of conformational restraints.

The next step was the introduction of the methylene acetal residue in **6** and **7** (Scheme 5). The initial plan was to effect this transformation in a single step, by exposure of the methoxymethyl derivatives **6** and **7** to dimethylboron bromide, following the conditions developed by Roush and co-workers *(38)*. However, this reaction was not successful for either substrate, presumably due to competing cleavage of the acetonide residue. More promising results were obtained with triol **29** (obtainable from controlled hydrolysis of **7**). Thus, treatment of **29** with dimethylboron bromide in the presence of di-t-butyl-4-methylpyridine at low temperature, led to **30** in 23% yield, together with the tetraol resulting from cleavage of the MOM ether. The reaction remains to be optimized. Application of these conditions to the corresponding triol from **6** was not successful. An alternative strategy was developed for this series. This plan arose from the finding that the MOM ether in **6** could be selectively removed (in the presence of the acetonide) under the dimethylboron bromide conditions. Thus the resulting diol **26** was treated with methylene dibromide and aqueous sodium hydroxide, following the conditions developed by Szarek *(39)*. This led to formation of the desired methylene acetal with concomitant removal of the silyl ether, in 71% yield. In preparation for alcohol protecting group processing, the silyl ether of the primary alcohol was reintroduced to give **28**. Diols **28** and **29** are primed for conversion to constrained sLe^x analogues **5** and **6** and related isoteres of sialic acid, through application of the dibutyltin oxide alkylation methodology (that was utilized in our earlier synthesis of **3**).

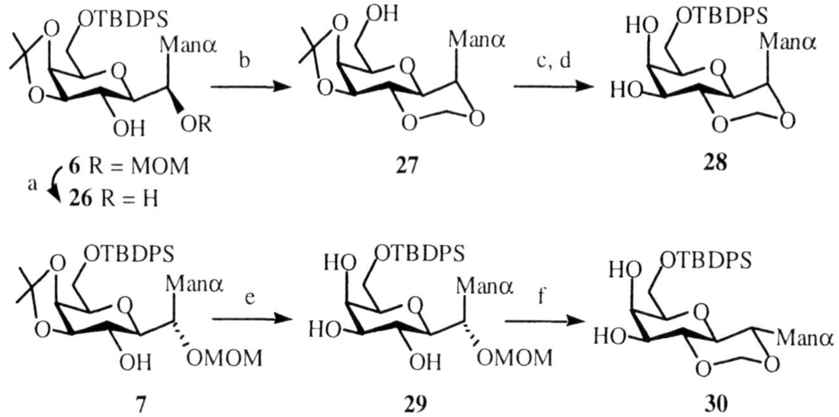

a: **6** R = MOM → **26** R = H; b → **27**; c, d → **28**

7 (e) → **29** (f) → **30**

(a) Me_2BBr, CH_2Cl_2 -78 °C -> rt; (b) CH_2Br_2, NaOH, 65 °C, 71%; (c) HCl, MeOH, 66%; (d) TBDPSCl, DMF, imidazole; (e) as in (c), 69%; (f) Me_2BBr, DTBMP, 4A mol. sieves, CH_2Cl2 -78 °C -> rt, **30** (23%) + tetrol (60%)

Scheme 5: Conformationally restrained C-glycosides

Structural analysis of C-glycoside products

The structures of **28** and **30** were confirmed by NMR analysis of their peracetate derivatives **31** and **32** (Fig. 2, Tables 1, 2). The stereochemistry of the

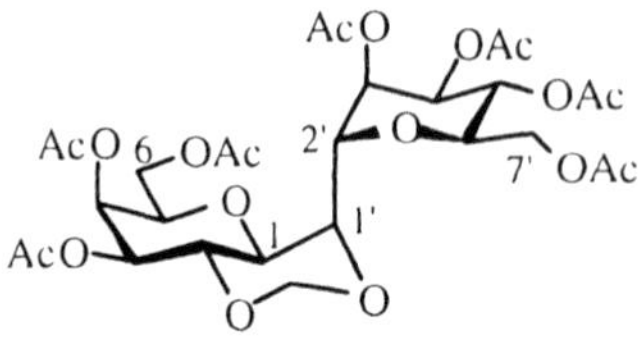

$J_{vicinal}$	**31**	**32**
$J_{1',1}$	6.5	9.8
$J_{1,2}$	10.0	9.8
$J_{2,3}$	10.0	9.8
$J_{3,4}$	3.0	3.0
$J_{4,5}$	0	0

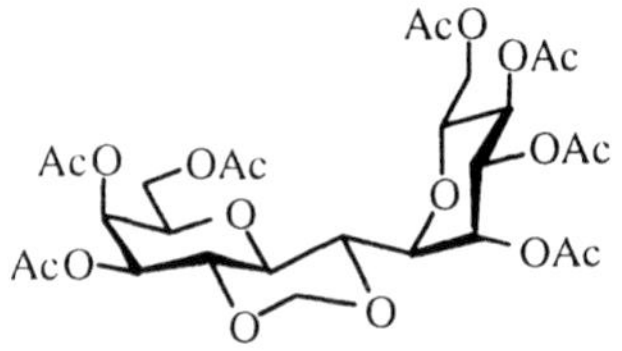

Fig. 2: Stereochemical analysis of peracetylated C-glycosides

aglycone segment and the configuration at the intersaccharide carbon were assigned on the basis of vicinal J values. Thus, $J_{1,2}$ = 10.0, $J_{2,3}$ = 10.0, $J_{3,4}$ = 3.0, $J_{4,5}$ = 0 Hz for **31**, and $J_{1,2}$ = 9.8, $J_{2,3}$ = 9.8, $J_{3,4}$ = 3.0, $J_{4,5}$ = 0 Hz for **32** are consistent with the 3,4-O-isopropylidene-β-C-galacto motif *(40)*. A $J_{1,1'}$ value of 9.8 Hz for **32** points strongly to an equatorial like attachment of the Man residue

Table 1: ^{1}H for C-glycosides **31** and **32**

Proton #	**31** / 1H ppm (J, Hz)	**32** / 1H ppm (J, Hz)
1'	4.19 (m)	3.77 (dd, 2.2, 9.8)
2'	4.56 (m)	4.46 (s)
3'	5.78 (dd, 3.0, 6.0)	5.84 (bs)
4'	5.70 (dd, 3.5, 6.0)	5.96 (dd, 4.0, 9.0)
5'	5.47 (t, 6.0)	5.72 (t, 9.0)
6'	4.07 (q, 5.7)	4.28 (m)
7'a	4.56 (m)	4.56 (dd, 5.0, 11.0)
7'b	4.56 (m)	4.28 (m)
1	3.47 (dd, 6.5, 10.0)	3.23 (t, 9.8)
2	4.51 (t, 10.0)	3.69 (t, 9.8)
3	5.19 (dd, 3.0, 10.0)	5.15 (dd, 3.0, 9.8)
4	5.54 (d, 3.0)	5.47 (d, 3.0)
5	3.33 (t, 6.0)	3.30 (t, 6.5)
6a	4.00 (dd, 6.0, 11.6)	3.99 (dd, 6.5, 10.8)
6b	4.19 (m)	4.10 (m)
O-CHa-O	5.06 (d, 6.0)	4.62 (d, 6.5)
O-CHb-O	4.71 (d, 6.0)	buried under m at 4.10
CH_3CO-	1.56, 1.65, 1.67, 1.68, 1.74, 1.75, 1.96	1.63, 1.68, 1.73, 1.74, 1.75, 1.78

Table 2: ^{13}CNMR for C-glycosides **31** and **32**

Carbon type	**31** ^{13}C NMR (ppm)	**32** ^{13}C NMR (ppm)
C$\underline{H}_3$CO:	20.7, 20.9, 21.0 (2C), 21.1 (2C)	20.8 (2C), 20.9 (2C), 21.0 (2C), 21.2
R$_3\underline{C}$O	61.8, 62.3, 67.4, 68.0, 68.1, 68.2, 68.7, 70.7, 71.1, 71.9, 72.6, 73.1, 75.5	61.7, 62.9, 67.3, 68.4 (2C), 69.9, 71.1, 72.8, 73.7, 74.6, 75.5, 75.7, 80.4
O-$\underline{C}$-O	90.5	93.5
CH$_3\underline{C}$O:	169.3, 169.7, 170.0 (2C), 170.2, 170.5 (2C)	169.7, 168.8 (2C), 169.9, 170.0, 170.3, 170.7

onto a chair-like dioxane ring. The corresponding J value for **31** (6.5 Hz), is somewhat larger than expected for equatorial-axial arrangement of vicinal protons on a chair-like dioxane. It appears that in this case the bulky pseudo-axial substituent results in a distorted, half-chair-like geometry, leading to the unusually large J value.

Summary

C-glycosides because of their chemical constitution (carbon vs. oxygen intersaccharide linker), allows for the construction of saccharide analogues that are restrained with respect to the pseudo glyconic and aglyconic torsions. Recognition probes based on such templates may provide less ambiguous structure activity information compared with more flexible structures. In this context, methylene acetals **28** and **30** are attractive scaffolds for the design of ligands for investigating the structural details of sLe^x-selectin recognition. The synthesis of C-glycosides of this type is particularly challenging when the aglycone segment is highly substituted (e.g. as in C-disaccharides). As illustrated here, the TIA based C-glycoside methodology provides a viable strategy to such complex frameworks.

Representative reaction procedures and physical data for C-glycoside derivatives

The experimental procedures and physical data for the reactions and products in the sequence from **14** -> **6** are typical for the C-glycosidation protocol.

General

Unless otherwise stated, all reactions were carried out under a nitrogen atmosphere in oven-dried glassware using standard syringe and septa technique. Chemical shifts are relative to the deuterated solvent peak or the tetramethylsilane (TMS) peak at (δ 0.00) and are in parts per million (ppm). The ^{1}H NMR assignments were determined from ^{1}H COSY experiments. High-resolution mass spectrometry (HRMS) data was obtained at the Mass Spectrometry Facility at University of Illinois, Urbana. Thin layer chromatography (TLC) was done on 0.25 mm thick precoated silica gel HF_{254} (Whatman) aluminum sheets. The chromatograms were observed under UV light and, or were visualized by heating plates that were dipped in ammonium molybdate solution. Unless otherwise stated, flash column chromatography (FCC) was performed using silica gel 60 (230-400 mesh) and employed a stepwise solvent polarity gradient, correlated with TLC mobility.

Physical data for the aglycone component: carboxylic acid 14

Colorless oil; R_f = 0.47 (10% MeOH/$CHCl_3$); ^{1}H NMR (300 MHz, $CDCl_3$) δ 3.49 (s, 3H), 3.74 (m, 3H), 3.91 (m, 2H), 4.01 (t, J = 3.7 Hz, 1H), 4.17 (m, 2H), 4.39 (m, 1H), 4.55-4.71 (m, 10H), 4.77 (d, J = 7.3 Hz, 1H), 4.85 (d, J = 6.9 Hz, 1H), 7.31-7.49 (m, 20H), 11.22 (s, 1H); ^{13}C NMR (75 MHz, $CDCl_3$) δ 56.5, 68.6, 70.8, 72.1, 72.2, 72.8, 73.2, 74.1, 74.4, 74.9, 75.7, 97.2, 127.4, 127.7, 127.8, 127.9, 128.3, 137.8, 137.9, 138.0, 138.2, 174.5; FABHRMS calcd for $C_{38}H_{42}O_9Na$ (M + Na) 665.2727, found 665.2722.

Physical data for the glycone component: TIA-alcohol 16

Colorless oil; R_f = 0.50 (10% EtOAc/petroleum ether); ^{1}HNMR (300 MHz, $CDCl_3$) δ 1.07 (s, 9H), 1.47, 1.49 (both s, 6H), 2.32 (br s, 1H, D_2O ex), 3.80 (m, 3H), 4.18 (dd, J = 2.0, 7.0 Hz, 1H), 5.44 (d, J = 7.0 Hz, 1H), 7.20-7.80 (m, 15H); ^{13}CNMR (75 MHz, $CDCl_3$) δ 26.3, 27.1, 27.5, 65.3, 70.1, 80.4, 85.4, 111.5, 127.6, 127.9, 129.1, 129.9, 132.0, 133.3, 134.0, 135.7. ESMS: 531 (M+Na); FABHRMS: calcd for $C_{23}H_{31}O_4Si$ (M-SC_6H_5) 399.1992, found 399.1992.

Thioacetal ester 22

DCC (324 mg, 1.58 mmol) was added at 0 °C to a solution of TIA-alcohol **16** (478 mg, 0.94 mmol), acid **14** (400mg, 0.63 mmol), and DMAP (16 mg, 0.13 mmol) in anhydrous benzene (50 mL). The reaction was warmed to rt and stirred for 2 h. The mixture was diluted with ether (10 mL) and filtered. The filtrate was washed with 0.1 N aqueous HCl and brine, dried (Na_2SO_4), filtered, and evaporated under reduced pressure. The residue was purified by FCC to give ester **22** (525 mg, 74% based on recovered alcohol): colorless oil; R_f = 0.32 (10% EtOAc/petroleum ether); IR (neat) 1759 cm^{-1}; ^1H NMR (300 MHz, $CDCl_3$) δ 1.16 (s, 9H), 1.47, 1.54 (both s, 3H ea), 3.38 (s, 3H), 3.62 (dd, J = 5.2, 9.4 Hz, 1H), 3.88-3.97 (m, 3H), 4.08-4.24 (m, 3H), 4.30-4.80 (m, 14H), 5.51 (m, 1H), 5.57 (d, J = 6.6 Hz, 1H), 7.34-7.49 (m, 29 H), 7.66 (m, 2H), 7.81 (m, 4H); ^{13}C NMR (75 MHz, $CDCl_3$) δ 26.5, 27.0, 27.4, 56.4, 62.5, 69.5, 71.2, 72.2, 72.6, 72.9, 73.2, 74.9, 75.8, 79.2, 84.8, 96.1, 111.6, 127.3-129.8 (several lines), 132.8, 133.0, 133.4, 135.6, 135.7, 138.2, 138.4, 138.4, 138.4, 138.8, 169.7; FABHRMS calcd for $C_{67}H_{76}O_{12}NaSiS$ (M + Na) 1155.4724, found 1155.4709.

Thioacetal-enol ether 23.

Tebbe reagent in THF (2.2 mL, 0.5 M, 1.17 mmol), was added dropwise under an atmosphere of argon at -78 °C, to a solution of ester **22** (315 mg, 0.28 mmol) and pyridine (0.1 mL) in anhydrous 3:1 toluene/THF (4 mL). The reaction mixture was warmed to rt, stirred at this temperature for 2 h, and then slowly poured into 1N aqueous NaOH at 0 °C. The resulting suspension was extracted with ether, and the organic phase was washed with brine, dried (Na_2SO_4), filtered, and concentrated under reduced pressure. The residue was purified by FCC on basic alumina (Brockmann I, 150 mesh) to give the enol ether **23** (219 mg, 83% based on recovered **22**): colorless oil; R_f = 0.56 (15% EtOAc/petroleum ether); ^1H NMR (500 MHz, C_6D_6) δ 1.13 (s, 9H), 1.45, 1.46 (both s, 3H ea), 3.15 (s, 3H), 3.82 (m, 5H), 3.91 (bs, 1H), 3.95 (dd, J = 4.5, 10.0 Hz, 1H), 4.05 (m, 1H), 4.12 (t, J = 9.5 Hz, 1H), 4.24 (d, J = 7.0 Hz, 1H), 4.34 (m, 3H), 4.42 (d, J = 12.0 Hz, 1H), 4.47 (d, J = 12.0 Hz, 1H), 4.51 (d, J = 12.0 Hz, 1H), 4.52-4.62 (m, 4H), 4.74 (m, 3H), 4.97 (d, J = 11.0 Hz, 1H), 5.92 (d, J = 7.5 Hz, 1H), 7.00-7.38 (m, 28H), 7.50 (d, J = 7.0 Hz, 2H), 7.71 (d, J = 7.0 Hz, 2H), 7.77 (m, 3H); ^{13}C NMR (75 MHz, C_6D_6) δ 27.1, 27.5, 28.0, 56.1, 62.1, 71.2, 72.1, 72.3, 74.2, 75.4, 75.6, 75.7, 75.8, 76.0, 76.3, 80.4, 80.7, 84.8, 88.0, 94.6, 112.1, 127.8-130.6 (several lines), 132.3, 133.6, 133.8, 135.2, 136.3, 139.5, 139.7, 139.9, 158.1; FABHRMS calcd for $C_{68}H_{78}O_{11}NaSiS$ (M + Na) 1153.4932 found 1153.4966.

C1-substituted glycal 8

A mixture of TIA-enol ether **23** (0.90 g, 0.80 mmol), 2,6-di-*tert*-butyl-4-methylpyridine (1.96 g, 9.56 mmol), and freshly activated, powdered 4A molecular sieves (1.00 g) in anhydrous CH_2Cl_2 (100 mL), was stirred for 15 min at rt under an argon atmosphere, then cooled to 0 °C. Methyl triflate (0.91 mL, 12.0 mmol) was then introduced and the mixture was warmed to rt and stirred for an additional 3 d, at which time triethylamine (1.67 mL) was added. The mixture was diluted with ether, washed with saturated aqueous $NaHCO_3$ and brine, dried (Na_2SO_4), filtered and evaporated under reduced pressure. The residue was purified by FCC over basic alumina (Brockmann I, 150 mesh) to give glycal **8** (550 mg, 82% based on **23**): clear oil; R_f = 0.60 (20% EtOAc/petroleum ether); ^1H NMR (300 MHz, C_6D_6) δ 1.16 (s, 9H), 1.33, 1.49 (both s, 3Hea), 3.28 (s, 3H), 3.78 (t, J = 6.5 Hz, 1H), 3.89-3.93 (m, 7H), 4.18 (m, 2H), 4.32-4.71 (m, 11H), 4.87 (m. 2H), 4.95 (d, J = 11.4 Hz, 1H), 7.12-7.39 (m, 24H), 7.45 (d, J = 7.0 Hz, 2H), 7.82 (m, 4H); ^{13}C NMR (75 MHz, C_6D_6) δ 27.4, 27.6, 29.2, 56.3, 63.9, 69.6, 71.0, 72.1, 72.4, 72.7, 73.5, 74.2, 75.1, 75.6, 76.3, 76.6, 80.8, 95.3, 102.8, 110.8, 127.8-130.5 (several lines), 134.0, 136.3, 136.4, 139.7, 139.8, 139.9, 151.9; FABHRMS calcd for $C_{62}H_{72}O_{11}NaSi$ (M + Na) 1043.4742, found 1043.4734.

β-C-galactoside 6

$BH_3.Me_2S$ (0.02 mL, 0.23 mmol) was added at 0 °C under an atmosphere of argon, to a solution of glycal **8** (24 mg, 0.0.02 mmol) in anhydrous THF (2 mL). The mixture was warmed to rt, stirred for an additional 1.5 h at this temperature, and cooled to 0 °C. The solution was then treated with a mixture of 3N NaOH (0.12 mL) and 30% aqueous H_2O_2 (0.05 mL) for 30 min. The mixture was diluted with ether, washed with saturated aqueous $NaHCO_3$ and brine, dried (Na_2SO_4), filtered and evaporated under reduced pressure. The residue was purified by FCC to give alcohol **6** (21 mg, 88%): colorless oil; R_f = 0.61 (30% EtOAc/petroleum ether); ^1H NMR (500 MHz, $CDCl_3$) δ 1.00 (s, 9H), 1.33, 1.44, (both s, 3H ea), 2.29 (d, J = 8.5 Hz, 1H, D_2O ex), 3.23 (s, 3H), 3.26 (m, 1H), 3.33 (bs, 1H), 3.43 (d, J = 8.8 Hz, 1H), 3.58-3.73 (m, 5H), 3.76 (d, J = 4.5 Hz, 1H), 3.80 (t, J = 8.0 Hz, 1H), 3.83 (dd, J = 2.0, 7.5 Hz, 1H), 3.90 (t, J = 9.5 Hz, 1H), 3.95 (dd, J = 2.0, 9.5 Hz, 1H), 4.14 (dd, J = 2.0, 7.5 Hz, 1H), 4.25 (m, 2H), 4.34 (d, J = 12.0 Hz, 1H), 4.46 (m, 4H), 4.53 (d, J = 6.0 Hz, 1H), 4.65 (d, J = 6.0 Hz, 1H), 4.72 (d, J = 6.0 Hz, 1H), 4.75 (d, J = 12.0 Hz, 1H), 7.13-7.31 (m, 26 H), 7.60 (4H); ^{13}C NMR (75 MHz, $CDCl_3$) δ 26.6, 27.0, 28.7, 56.0, 62.4, 69.3, 69.9, 71.8, 72.4, 73.1, 73.4, 74.5, 74.7, 75.6, 75.9, 76.0, 76.8, 78.2, 78.6, 78.8, 98.5, 109.3, 127.4-129.8 (several lines), 133.1, 133.5, 135.5, 135.7, 138.3,

138.4, 138.8; FABHRMS calcd for $C_{62}H_{74}O_{12}NaSi$ (M + Na) 1061.4847, found 1061.4840.

Acknowledgment

This investigation was supported by NIH grant GM 57865. "Research Centers in Minority Institutions" award RR-03037 from the National Center for Research Resources of the NIH, which supports the infrastructure {and instrumentation} of the Chemistry Department at Hunter, is also acknowledged. The contents are solely the responsibility of the authors and do not necessarily represent the official views of the NCRR/NIH.

References

1. Vestweber, D.; Blanks, J. E. *Physiol. Rev.* **1999**, *79*, 181–213.
2. Springer, T. A. *Cell* **1994**, *76*, 301-314.
3. Lasky, L. A. *Science* **1992**, *258*, 964–969.
4. Paulson, J. C. in Adhesion: Its role in inflammatory diseases; Harlan, J. M., Liu, D. Y., Eds.; W. H. Freeman: New York 1992, p 19–42.
5. Boschelli, D.H. *Drugs of the Future* **1995**, *20*, 805-816.
6. Mousa, S. A. *Drugs of the Future* **1996**, *21*, 283-289.
7. Simanek, E.E.; McGarvey, G.J.; Jablonowski, J.A.; Wong, C.-H. *Chem. Rev.* **1998**, *98*, 833-862.
8. Roy, R. in *Carbohydrates in Drug Design*; Witczak Z. J., Nieforth, K. A., Eds.; Marcel Dekker: New York, 1997; p 83-135.
9. Musser, J. H.; Anderson, M. B.; Levy, D. E. *Curr. Pharm.Design* **1995,** *1*, 221-232.
10. Selected recent studies: Chervin, S. M.; Lowe, J. B.; Koreeda, M. *J. Org. Chem.* **2002**, *67*, 5654-5662 and ref. 11-13
11. Hanessian, S.; Mascitti, V.; Rogel, O. *J. Org. Chem.* **2002**, *67*, 3346-3354.
12. Thoma, G.; Magnani,; J. L.; Patton, J. T.; Ernst, B.; Jahnke, W. *Angew. Chem. Int. Ed.* **2001**, *40*, 1941-1945.
13. Thoma, G.; Banteli, R.; Jahnke, W.; Magnani, J. L.; Patton, J. T. *Angew. Chem. Int. Ed.* **2001**, *40*, 3644-3647.
14. Hiruma, K.; Kajimoto, T.; Weitz-Schmidt, G.; Ollmann, I.; Wong, C.-H. *J. Am. Chem. Soc.* **1996**, *118*, 9265-9270.
15. Shibata, K.; Hiruma, K.; Kanie, O.; Wong, C.-H. *J. Org. Chem.* **2000**, *65*, 2393-2398.

16. Scheffler, K.; Ernst, B.; Katopodis, A.; Magnani, J. L.; Wang, W. T.; Weisemann, R.; Peters, T. *Angew. Chem. Int. Ed.* Engl. **1995,** *34*, 1841-1844.
17. Hensley, P.; McDevitt, P. J.; Brooks, I.; Trill, J. J.; Field, J. A.; McNulty, D. E.; Connor, J. R.; Griswold, D. E.; Kumar, N. V.; Kopple, K. D.; Carr, S. A.; Dalton, B. J.; Johanson, K. *J. Biol. Chem.* **1994**, *269*, 23949-23958.
18. Cooke, R. M.; Hale, R. S.; Lister, S. G.; Shah, G.; Weir, M. P.; *Biochemistry* **1994**, *33*, 10591-10596.
19. Somers, W. S.; Tang, J.; Shaw, G. D.; Camphausen, T. *Cell* **2000**, *103*, 467–479.
20. Poppe, L.; Brown, G. S.; Philo, J. S.; Nikrad, P. V.; Shah, B.H. *J. Am. Chem. Soc.* **1997**, *119*, 1727–1736.
21. Hiramatsu, Y.; Tsujishita, H.; Kondo, H. *J. Med. Chem.* **1996**, *39*, 4547–4553.
22. Bertozzi, C. R. *Chem. Biol.* **1995**, 703–708.
23. Kogan, T. P.; Revelle, B. M.; Tapp, S.; Beck, P. J.; *J. Biol. Chem.* **1995**, *270*, 14047–14055.
24. Cheng, X.; Khan, N.; Mootoo, D. R. *J. Org. Chem.* **2000**, *65*, 2544-2547.
25. Asensio, J. L.; Cañada, F. J.; Cheng, X.; Khan, N.; Mootoo, D. R.; Jiménez-Barbero, J. *Chem. Eur. J.* **2000**, *6*, 1035-1041.
26. For examples of conformationally restrained carbohydrate ligands: Bundle, D. R.; Alibés, R.; Nilar, S.; Otter, A.; Warwas, M.; Zhang, P. *J. Am. Chem. Soc.* **1998**, *120*, 5317-5318 and ref. 27.
27. Navarre, N.; Amiot, N.; Van Oijen, A.; Imberty, A.; Poveda, A.; Barbero, J. J-.; Cooper, A.; Nutley, M. A.; Boons, G-. J. *Chem. Eur. J.* **1999**, *5*, 2281-2294.
28. Khan, N.; Cheng, X.; Mootoo, D. R. *J. Am. Chem. Soc.* **1999**, *121*, 4918-4919.
29. For a conceptually similar C-glycoside synthesis based on ring closing olefin metathesis: Liu, L.; Postema, M. H. D. *J. Am. Chem. Soc.* **2001**, *123*, 8602-8603.
30. Reviews on C-glycoside synthesis: Postema, M. H. D. *C-Glycoside Synthesis*, CRC Press, Boca Raton, Fl, 1995 and ref. 31-34.
31. Levy, D.; Tang, C. *The Synthesis of C-Glycosides*, Pergamon, Oxford, 1995.
32. Beau, J.-M.; Gallagher, T. *Topics Curr. Chem.* **1997**, *187*, 1-54.
33. Togo, H.; He, W.; Waki, Y.; Yokoyama, M. *Synlett* **1998**, 700-717.
34. Du, Y.; Linhardt, R. J.; Vlahov, J. R. *Tetrahedron* **1998**, *54*, 9913-9959.
35. De Armas, P.; Francisco, C. G.; Suarez, E. *Angew. Chem. Int. Ed. Engl.* **1992**, *31*, 772-774.
36. Neises, B.; Steglich, W. *Angew. Chem. Int. Ed. Engl.* **1978**, *17*, 522-523.

37. Pine, S. H.; Pettit, R. J.; Geib, G. D.; Cruz, S. G.; Gallego, C. H.; Tijerina, T.; Pine, R.D. *J. Org. Chem.* **1985**, *50*, 1212-1218.
38. Powell, N. A.; Roush, W. R.; *Org. Lett.* **2001**, *3*, 453-456.
39. Kim, K. S.; Szarek, W. A. *Synthesis* **1978**, 48-50.
40. Barili, P. L.; Catelani, G.; D'Andrea, F.; Mastrorilli, E.; *J. Carbohydr. Chem.* **1997**, *16*, 1001-1010.

Chapter 8

Effects of Linker Rigidity and Orientation of Mannoside Clusters for Multivalent Interactions with Proteins

René, Roy[1,2,*], M. Corazon Trono[2], and Denis Giguère[1]

[1]Current addres: Department of Chemistry, Université du Québec à Montréal, P.O. Box 8888, Succ. Centre-Ville, Montréal, Québec H3C 3P8, Canada
[2]Department of Chemistry, University of Ottawa, Ottawa, Ontario K1N 6N5, Canada

A series of di-, tri-, and hexameric α-D-mannopyranoside clusters having rigid linkers were constructed in order to evaluate the effects of inter-saccharide distances upon multivalent binding interactions with the phytoheamagglutinin from *Canavalia ensiformis* (Concanavalin A). To this end, O-aryl and O-propargyl glycosides were synthesized and then coupled together using transition metal catalyzed cross-coupling reactions. Thus, ortho-, *meta*-, and *para*-iodophenyl mannosides were treated with acetylene or propargyl mannoside under palladium(0)-catalyzed Sonogashira conditions to afford mixtures of cross-coupled symmetrical and unsymmetrical dimers. Dicobalt octacarbonyl benz-annulation of acetylenic mannosides provided hexamers showing potent cross-linking ability.

Introduction

Multivalent carbohydrate protein interactions are ubiquitous in biological systems (*1*). It is believed that nature uses this strategy to compensate for the generally low affinity and poor selectivity of carbohydrate ligands. There has been several reports demonstrating that glycopolymers (2-5), glycodendrimers (*6-8*), and even small glycoclusters (*9*), could achieve enhanced avidity through various reaction mechanisms involving for example "wrapping around" virus particles (*5*), chelate effects (*10*), and cross-linking effects (*11*). The last situation has been particularly well documented for soluble proteins possessing multiple carbohydrate recognition domains (CRDs). Interestingly, several proteins have homologous or very analogous amino acid sequences in their CRDs. This is the case for examples for the family of galectins, mannose binding proteins, and sialoadhesins that bind to their respective carbohydrate ligands with more or less similar affinity constants.

Moreover, Sharon (*12*) has already demonstrated that bacterial lectins at the tip of fimbriae possess subtle but different affinities for aromatic α-D-mannopyranosides, thus illustrating the effect of what we coined "sub-site-assisted aglycone binding" (*13*). For instance, α-D-mannopyranosides **2-4** (Scheme 1) showed 69 to 1015-fold increased inhibitory properties in comparison to methyl mannoside **1** (*12*). Additionally, discrete differences can be observed between the inhibitory properties of a given mannoside derivative against different strains of bacteria, thus illustrating again that selectivity can be achieved.

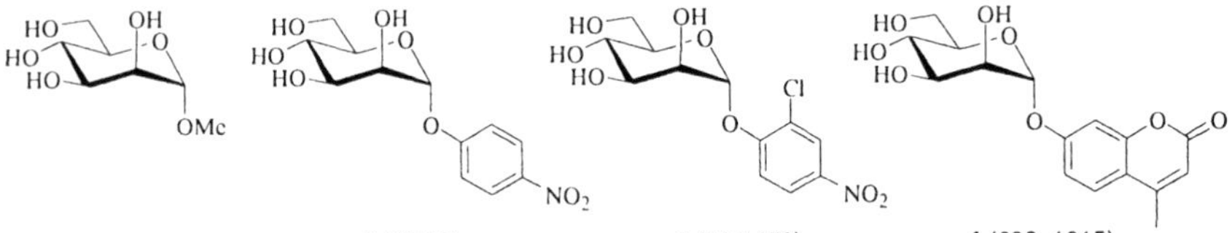

1 (1.0) **2** (69,70) **3** (717,470) **4** (600, 1015)

Scheme 1. Different mannoside derivatives and their relative inhibitory properties against the interaction of fimbriated E. coli O25 or O128 in agglutination of yeasts or adherence to epithelial cells, respectively.

The above situation has also been exemplified with recent findings in the galectin field, wherein lactose and N-acetyllactosamine derivatives **5** and **6** were showed to be 14x and 50x times more potent than their parent saccharides against galectin-3, respectively (Scheme 2). In the first case (*14*), compound **5** has its aglycone modified by an aromatic residue, while compound **6** (*15*), screened from a large library of 3'-derivatives, possesses an aromatic amide functionality that was postulated to occupy an hydrophobic pocket in the active site of galectin-3. The efficacy of lactoside **5** was further increased by the development of wedge-type glycodendrimers (*14*). Indeed, a tetravalent structure derived from **5** was found to be 1071x more potent than the monovalent lactoside in a solid-phase competitive inhibition assay using asialofetuin (ASF) as coating antigen and biotin-labeled galectin-3 for detection.

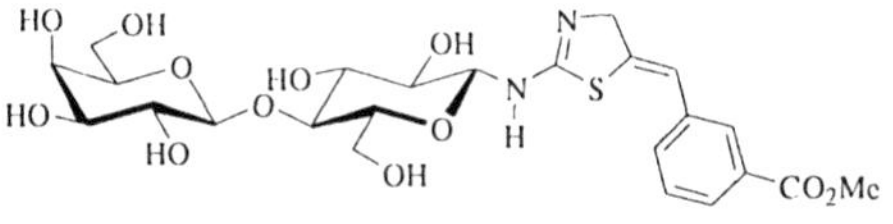

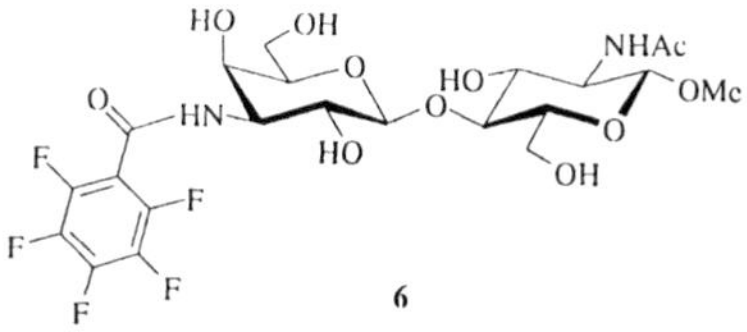

Scheme 2. Ligands used for the selective inhibition of galectin-3.

Using galactoside clusters having similar aromatic aglycones to those described herein for the mannosides, we found that a trimeric structure had the highest affinity for galectin-3 while a tetramer was more selective for galectins-1 and 5 (*16*). These results point toward a combine approach for the design of more potent and selective carbohydrate mimetics.

Another example of selective inhibitions was observed within the sialic acid binding immunoglobulin (Ig)-like lectins (Siglecs) that comprise a family of 11 type I membrane receptors involved in cell-cell recognition events. The synthesis of various sialoside analogs differing by their substituents at position 9 of the glycerol side chain illustrated again that compounds with significantly enhanced affinities and specificities could be achieved (*17*). Scheme 3 below shows the structural differences between compounds **7** and **8** in which the terminal C-9 alcohol functionality was exchanged for a benzamide residue. Compound **8** was 790x more potent than **7** in its binding affinity for the myelin-associated glycoprotein (MAG), known to play an important role in maintaining myelin-axon integrity. The additional benzamide residue was taught to be in closer contact with a tyrosine residue in the vicinity of the active site.

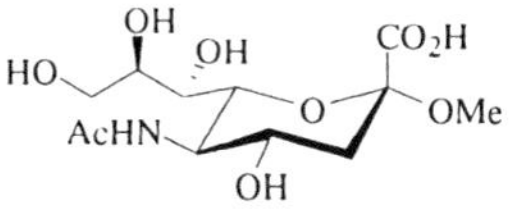

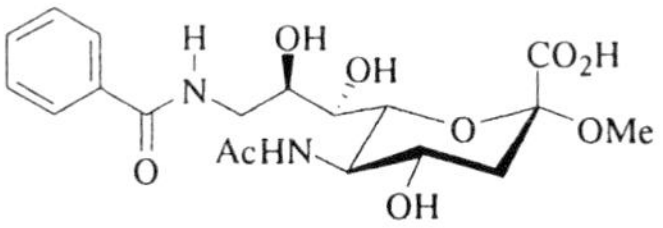

*Scheme 3. Structures of sialoside derivatives showing selective inhibitory potencies against a family of type 1 Siglec membrane receptors. Compound 8 was 790 fold more potent than **7** against MAG (Siglec-4).*

Discussion

Proteins having mannoside recognition domains are also ubiquitous and are found within a wide range of mammalian cells and microorganisms (*18*). Several of these mannose binding proteins have to discriminate self from non-self mannose-containing glycoconjugate epitopes for proper handling of the biochemical information. For instance, macrophages, dendritic cells, and hepatocytes are involved in immune regulations and glycoprotein uptake. They can also recognize the high mannose glycoprotein of HIV (*19*) and other microorganisms (*12*). Paradoxally, pathogens such as *E. coli* and *Pseudomonas aeruginosa* through their fimbriated lectins, can also bind to multiantennary mannosides on mammalian cell surfaces (*12*). Thus, it is highly desirable to gain insight into the nature, number, and topology of mannoside clusters required for a specific recognition event.

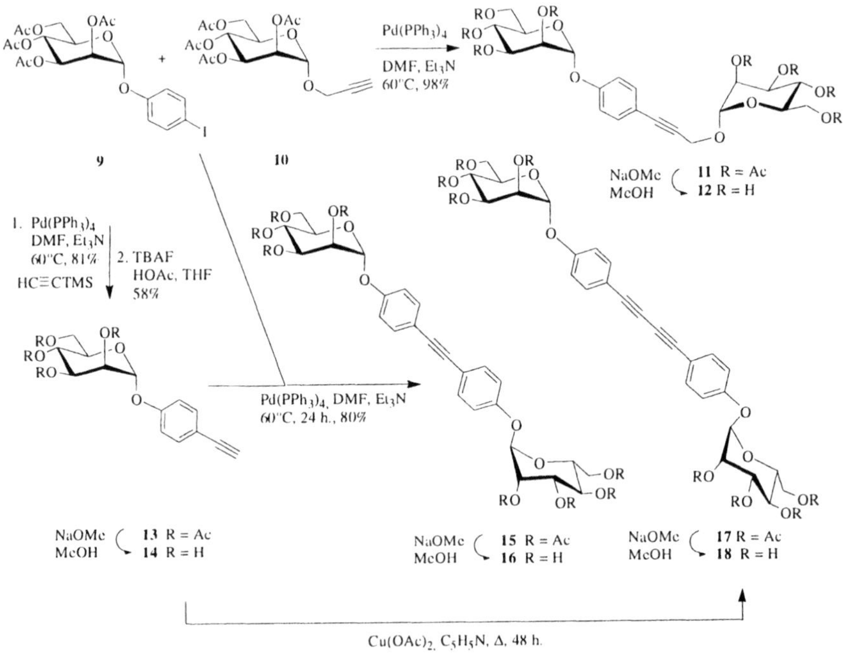

Scheme 4. Palladium(0)-catalyzed syntheses of mannoside dimers using Sonogashira cross-coupling reactions.

In our continued efforts (*20-22*) to prepare mannoside ligands of improved binding properties against mannose binding proteins (MBP) and the like, we describe herein the syntheses of mannoside dimers, trimers, and hexamers having rigid aromatic aglycones that were built using palladium(0)-catalyzed cross-coupling Sonogashira reactions (*23*). To this end, key *orto-*, *meta-*, and *para-*iodophenyl α-D-mannopyranosides were prepared from the corresponding arenols using mannose pentaacetate and boron trifluoride etherate in methylene chloride to afford mannosides **9** (*24*), **19**, and **21** in 81%, 36%, and 63% yield respectively (Scheme 4 and 5). The known (*24, 25*) propargyl mannoside **10** was also prepared (78% yield) by a Lewis acid-catalyzed reaction between mannose pentaacetate (**32**) and propargyl alcohol.

To initially test the optimum catalyst and cross-coupling conditions, treatment of *p*-iodophenyl mannoside **9** and propargyl mannoside **10** was attempted with tetrakis(triphenylphosphine)palladium(0) in DMF and triethylamine at 60°C to provide unsymmetrical dimer **11** in 98% yield (Scheme 3) (*26*). Quantitative de-*O*-acetylation under Zemplén conditions (NaOMe, MeOH) afforded dimer **12** in essentially quantitative yield. Then, treatment of **9** with (trimethylsilyl)acetylene, followed by removal of the silyl-protecting group under buffered conditions (HOAc) in THF afforded *p*-ethynylphenyl α-D-mannopyranoside **13** in 58% overall yield which after Zemplén treatment gave **14** quantitatively. Analogous handling on **19** and **21** gave **20** in only 10% yield and **22** in 55% yield (Scheme 4). An improved preparation of the *ortho* isomer **20** was achieved using dichlorobis(triphenylphosphine)palladium(II) ($PdCl_2(PPh_3)_2$) in the presence of triphenylphosphine, copper(I) iodide and acetylene gas bubbled into the reaction mixture (DMF, Et_3N, 60°C, 44%).

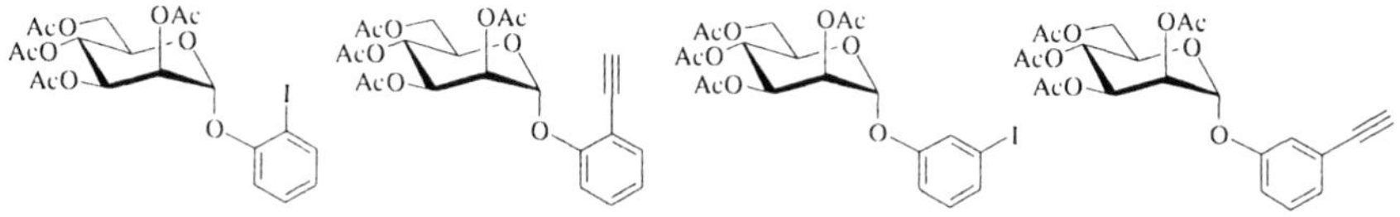

Scheme 5. Structures of the ortho-and meta-aryl mannosides used in the cross-coupling reactions.

Homodimerization of aryl iodides (**9, 19, 21**) with an acetylene bridge was then conducted at room temperature ($PdCl_2(PPh_3)_2$, CuI, Et_3N, 24 h.) to provide symmetrical dimer **15** (58%) accompanied by the acetylenic monomer **13** (27%). Attempts to provide the corresponding *ortho* and *meta* monoacetylene-bridged dimers **23** and **26** were troublesome since they were invariably contaminated with doubly bridged-dimers (**25**, **28**, Scheme 5) as inseparable mixtures (2:1 and

1:1.4 ratios, respectively). These compounds were obtained directly from **20** and **22** following an oxidative homocoupling in the presence of the copper catalyst. The situation was however remedied by using a cross-coupling strategy between the iodides and the ethynylphenyl mannosides in the absence of copper co-catalyst. Thus, treatment of **9** and **13**, **19** and **20**, and **21** and **22** ($Pd(PPh_3)_4$, Et_3N, DMF, 60°C, 24 h.) afforded the corresponding dimers after Zemplén deprotection (**16** (80%), **23** (100%), and **26** (100%)) as single products.

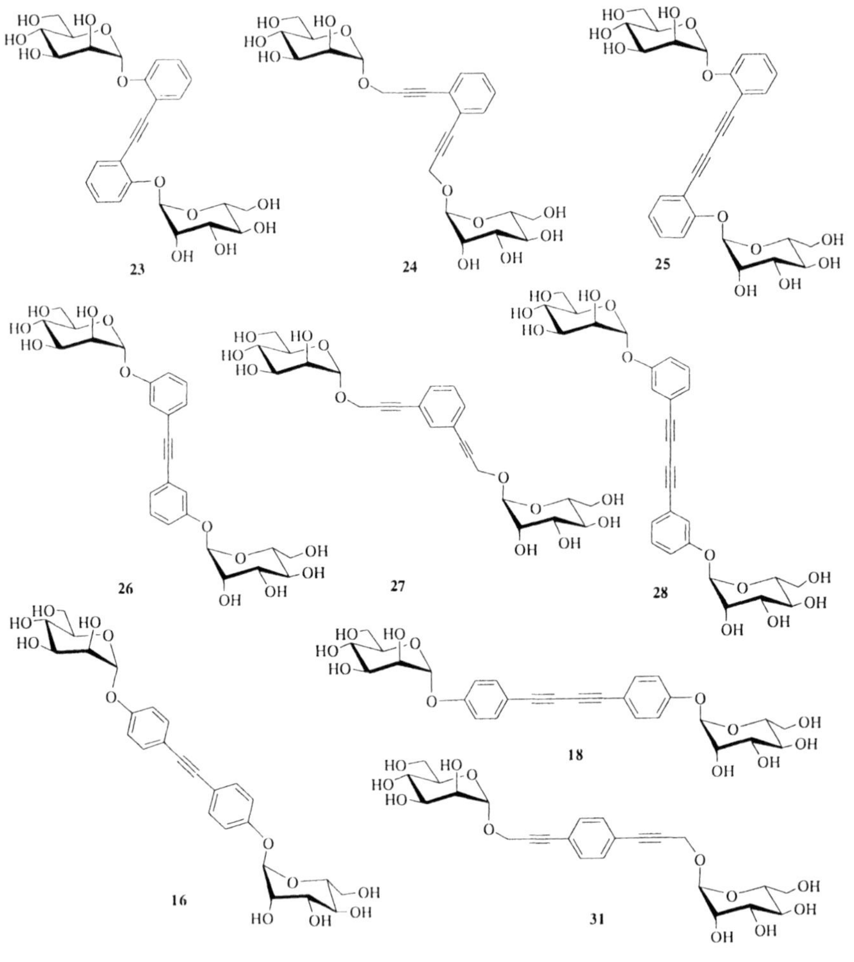

Scheme 6. Comparative structures of mannoside dimers prepared in this study.

Diacetylene-bridged dimers **18**, **25**, and **28** were more conveniently prepared (47-68% yields after Zemplén treatment) by a modified Glaser (*27*) oxidative homocoupling under Hay's conditions (*28*) of the corresponding ethynylphenyl mannosides using copper(II) acetate in refluxing pyridine (48 h.).

The syntheses of other dimers are also illustrated in Scheme 6. Thus, treatment of 2-propynyl α-D-mannopyranoside **10** with diiodobenzene (*para* isomer illustrated) under the optimum Sonogashira conditions described above provided **30** that after deprotection gave **31** in quantitative yield. For comparison purpose, dimer **36** and **38** (*29*) were prepared from peracetate **32** using bisphenol A (**33**) and 2-butyne-1,4-diol (**34**) under Lewis acid conditions (Scheme 7).

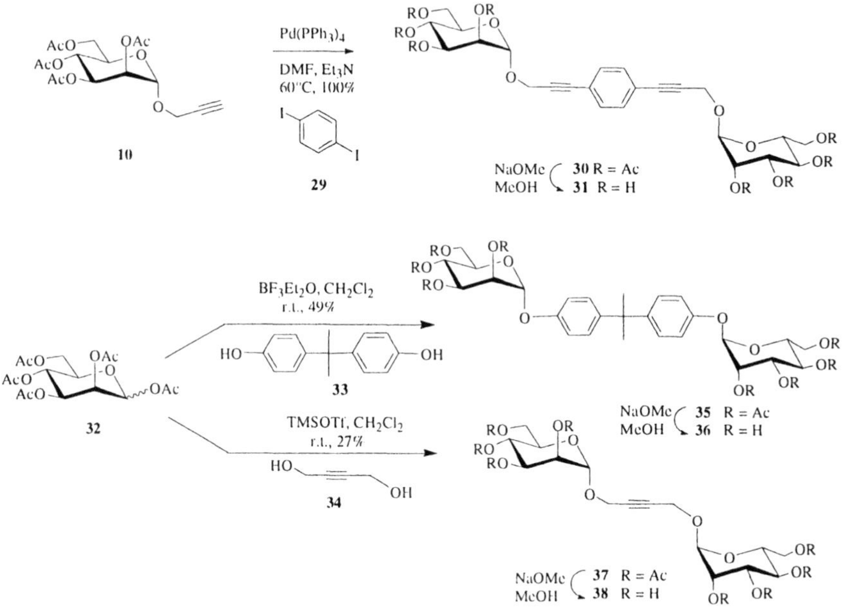

Scheme 7. Synthesis of other mannosides dimers using a tethering unit.

The human mannose-binding protein (hMBP) is a member of the collectin family of molecules that play a role in first line host defense. Found in the serum of many mammalian species (*30*), this protein mediates immunoglobin-independent defensive reactions against pathogens. Its action occurs within minutes of exposure to an infectious agent providing immediate defense during the 1-3 days lag period required for induction of specific antibodies. The mannose-binding proteins function by recognizing oligo-mannose on the cell surface of various bacteria and viruses. They bind and neutralize them by complement-mediated cell lysis or facilitate their recognition by phagocytes.

Under unclear physiological conditions, the monomeric structure of the hMBP forms clusters of 18 identical subunits arranged as hexamer of trimers, adopting a bouquet-like structure (*31*). Each subunit chain consists of four distinct regions. The oligomerization is possible through the disulfide bonds in the cysteine-rich region and due to the characteristic properties of the collagen-like domain and the neck region. The clustering of three CRDs in close proximity significantly increases the affinity for glycoconjugates present on pathogens. In order to better understand to role and selectivity of this carbohydrate-binding protein in comparison to other mannoside-active proteins, **it became of interest to synthesize other clusters composed of three and six α-D-mannopyranoside residues (Scheme 8 and 9).**

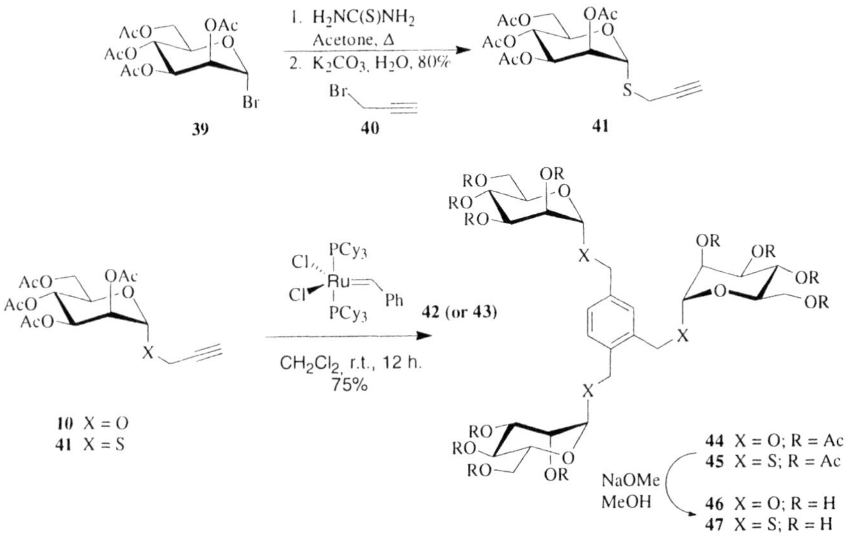

1. $H_2NC(S)NH_2$ Acetone, Δ
2. K_2CO_3, H_2O, 80%

39 40 41

CH_2Cl_2, r.t., 12 h. 75%

42 (or 43)

10 X = O
41 X = S

NaOMe MeOH

44 X = O; R = Ac
45 X = S; R = Ac
46 X = O; R = H
47 X = S; R = H

Scheme 8. Single step synthesis of mannoside trimers using a novel Grubb's catalyzed benzannulation reaction of O- and S-propargyl mannopyranosides.

Benzannulation of three alkynes is a well-known process that is occurring in the presence of dicobalt octacarbonyl ($Co_2(CO)_8$) (*29, 32-34*). To this end, we initially prepared O- and S-mannoside **10** and **41**, the latter being prepared from acetobromomannose **39** under standard conditions (*35*). Because the Grubbs' catalyst **42** or its second generation **43** (*36*) was known to be involved in ene-yne cross metathesis reactions (*37*), we had originally attempted to prepare mannoside analogs by cross-coupling **10** with various alkene derivatives. To our surprise, instead of reacting with its alkene partners, **10** preferred to react three consecutive times with itself to give good yield of a mixture (9:1) of regioisomeric benzannulated products, the major unsymmetrical isomer **44** being illustrated in Scheme 7 (*38*). This originally unexpected event was clearly the result of a [2+2+2] alkyne cyclotrimerization that is well-known to proceed with $Co_2(CO)_8$. The same protocol was also attempted on the thioglycoside **41** to afford **45**. Acetate deprotection (NaOMe, MeOH) afforded water-soluble trimers **46** and **47**.

To further expand the family of mannoside clusters obtainable from the alkyne derivatives, we also prepared cyclotrimerized hexamers using dicobalt octacarbonyl according to a known strategy (*34*). Thus dimer **37** and **30** were treated with $Co_2(CO)_8$ in refluxing dioxane to provide **48** and **50** in 61% and 42% yield, respectively (Scheme 9).

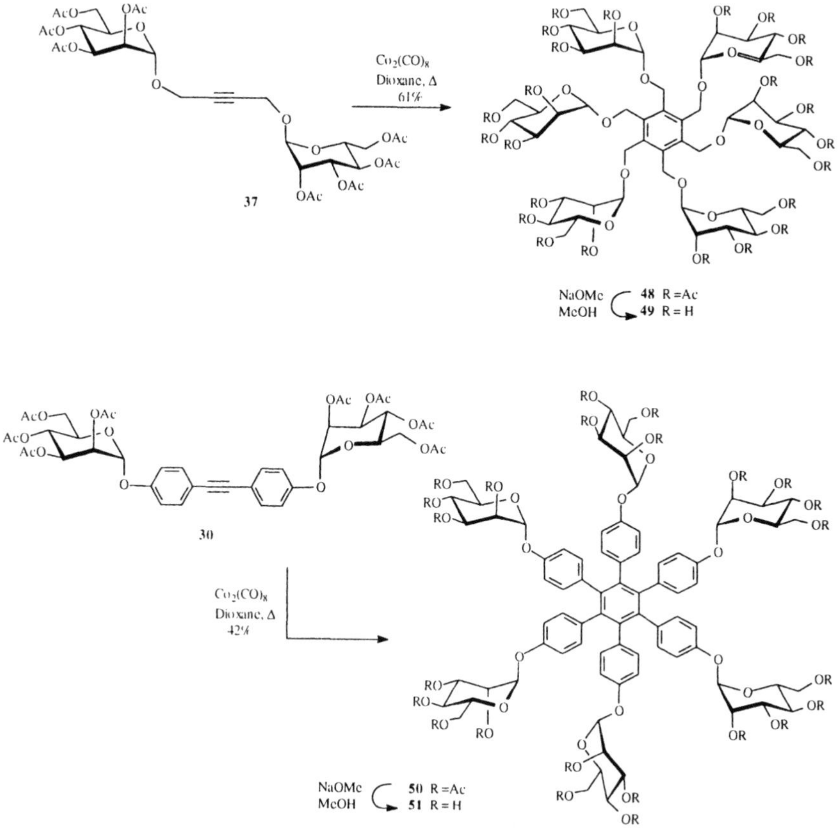

*Scheme 9. Alkyne cyclotrimerizations of **37** and **30** catalyzed by dicobalt octacarbonyl.*

After Zemplén treatment under usual conditions, the sparingly water-soluble hexamers **49** and **51** were obtained in quantitative yields. Their modest solubility in water has however not jeopardized their biological evaluations in solid-phase competitive inhibition assays as well as in their cross-linking abilities with a model plant lectin, Concanavalin A, known to form cross-linked lattices in the presence of multiantennary glycans (*39*).

Cross-linking assays

The relative ability of the above mannoside clusters to bind efficiently to a multivalent protein was evaluated using an adaptation of nephelometric assays (turbidimetry) (*40*). The tetrameric phytoheamagglutinin from *Canavalia ensiformis* (Concanavalin A) was used as a model since its cross-linking ability in the presence of mannoside clusters is well-documented (*39*). Using microtiter plates (300μL), incremental amount of the clusters were added to a fixed quantity of the lectin in phosphate buffer saline (PBS) and the turbidity was checked at time interval using absorbances at 490nm.The initial results were plotted as maximum binding as a function of time and indicated that dimer **16** was superior to trimer **46**. Trimer **49** failed to form a cross-linked lattice and hexamer **51** showed the best and fastest cross-linking ability. Using the best candidate **51** as standard normalized to 100% binding, the other clusters were similarly evaluated. The results are shown in Fig. 1.

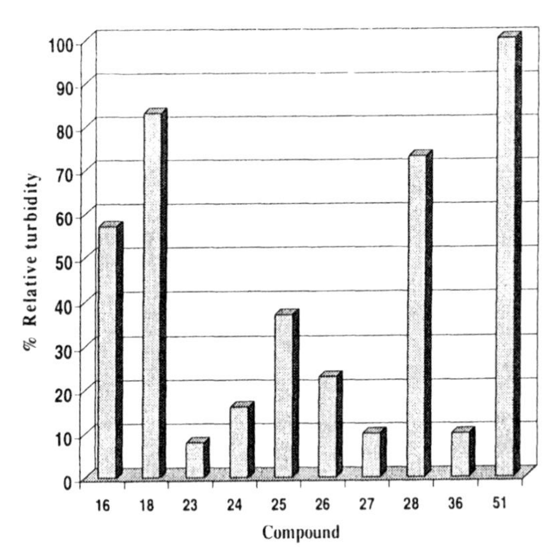

*Fig. 1. Relative cross-linking ability of α-D-mannopyranoside clusters toward tetrameric Con A. Measurements were done in microtiter plates at 490 nm after 5 min. and standardized to the maximum binding of hexamer **51** at that time.*

While most compounds could form insoluble cross-linked lattices very rapidly (except trimer **49**), dimers **18** and **28**, having the longest inter-mannoside distances showed to be the best candidates within the dimeric family of derivatives. For instance, dimer **18** was capable of precipitating 80% of the maximum value reached by the best candidate, hexamer **51**, after only 5 min. Interestingly, most *ortho-* and some *meta*-substituted dimers, having short distances between the sugar residues, were the slowest to form stable insoluble lattices. It is however noteworthy to mention that after prolonged reaction time (up to 70 min.), most candidates succeeded in attaining constant absorbance, albeit with lower values, therefore indicating that the cross-linking phenomena are under dynamic equilibrium processes and that some clusters could organize the lattice at faster rates than other analogs.

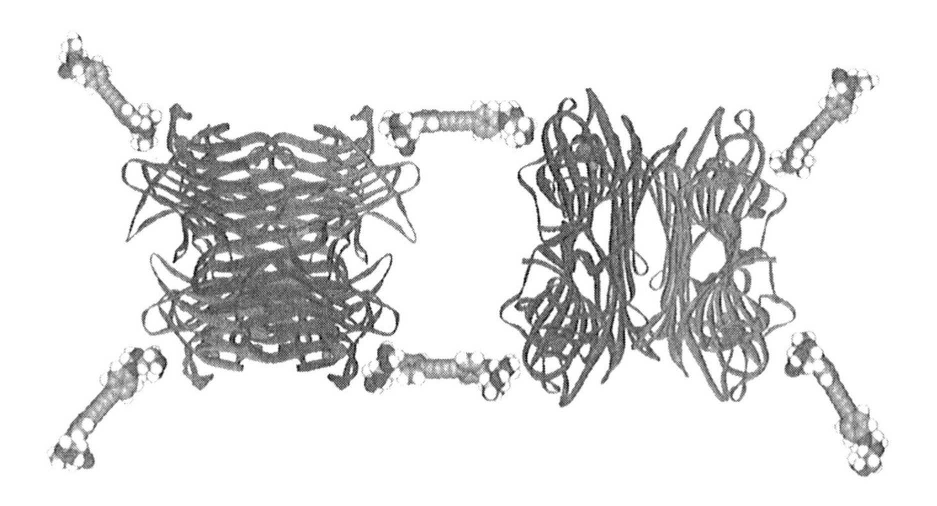

*Fig. 2. Schematic representation of Concanavalin A lectin tetramer forming a cross-linked lattice with dimmer **18**. (See color plate 1.)*

A schematic representation of the cross-linked behavior of the best dimeric candidate (**18**) is illustrated (Fig. 2). The hypothetical lattice structure shown was based on the general behavior of natural (*39*) as well as non-natural (*41*) mannosides. The above cross-linking process is suggested to be analogous to the one used by nature to provide high selectivity and affinity within, otherwise similar carbohydrate ligands.

Experimental

Hexakis-(2,3,4,6-tetra-*O*-acetyl-α-D-mannopyranosyloxy)hexaphenylbenzene (50). A solution of bis-*para*-(2,3,4,6-tetra-*O*-acetyl-α-D-mannopyranosyloxy)diphenyl-acetylene (**30**) (0.35 g, 0.40 mmol) in 5 mL freshly distilled 1,4-dioxane was refluxed under a stream of N_2 gas. After 5 minutes of reflux, dicobalt octacarbonyl ($Co_2(CO)_8$) (0.019 g, 0.056 mmol) was added. The reaction mixture was refluxed for a further 24 hours, after which time, the reaction was judged complete by TLC on silica gel (ethyl acetate/ toluene 6.4: 1.5, v:v). The solvent was evaporated under reduced pressure and the resulting residue was purified by column chromatography on silica gel using hexane/ ethyl acetate 1:4 as eluent. The cyclotrimerized product was obtained as yellow crystalline powder (0.14 g, 42 % yield); m.p. 85 °C; IR 2940, 1750, 1435, 1371, 1226 cm^{-1}; ^1H NMR ($CDCl_3$, 500 MHz): δ (ppm) 6.62 (C_6H_4, d, J=8.7 Hz, 12H), 6.58 (C_6H_4, d, J=8.5 Hz, 12H), 5.40 (H-3), dd, $J_{2,3}$=3.7 Hz, $J_{3,4}$=10.0 Hz, 6H), 5.26 (H-2, H-4, dd, J=9.7 Hz, J=10.7 Hz, 12H), 5.24 (H-1, s, 6H), 4.20 (H-6a, dd, $J_{5,6a}$=5.2 Hz, $J_{6a,6b}$=12.5 Hz, 6H), 3.93 (H-5, H-6b, m, 12H), 2.10, 2.00, 1.99, 1.98, 1.94 (2x) ($COCH_3$, 72H); ^{13}C NMR ($CDCl_3$, 125.7 MHz): δ (ppm) 170.3, 169.8, 169.7, 169.7, (C=O), 153.3, 139.9, 135.3, 132.3, 115.2 (C_6H_4), 96.0 (C-1), 69.4, 69.2, 69.0, 68.8 (C-2, C-3, C-5), 65.8 (C-4), 20.9, 20.7, 20.6 ($COCH_3$); FAB-MS $[M+K]^+$ *m/z* (rel. intensity %) calcd for $C_{126}H_{138}O_{60}$: 2649.74; found: 2650.70 (0.1).

Acknowledgements: We are thankful to the Natural Sciences and Engineering Research Council of Canada (NSERC) for partial support of this work. We also thank Dr. S. K. Das for useful suggestions and assistance.

References

1. Varki, A. *Glycobiology* **1993,** *3*, 97-130.
2. Bovin, N. V.; Gabius, H.-J. *Chem. Soc. Rev.* **1995,** *24*, 413-421.
3. Roy, R. *Trends Glycosci. Glycotechnol.* **1996,** *8*, 79-99.
4. Kiessling, L. L.; Pohl, N. L. *Chem. Biol.* **1996,** *3*, 71-77.
5. Mammen, M.; Choi, S. K.; Whitesides, G. M. *Angew. Chem. Int. Ed.* **1998,** *37*, 2754-2794.
6. Roy, R. *Trends Glycosci. Glycotechnol.* **2003,** *15*, 291-310.
7. Röckendorf, N.; Lindhorst, T. K. *Topics Curr. Chem.* **2001,** *217*, 201-238.

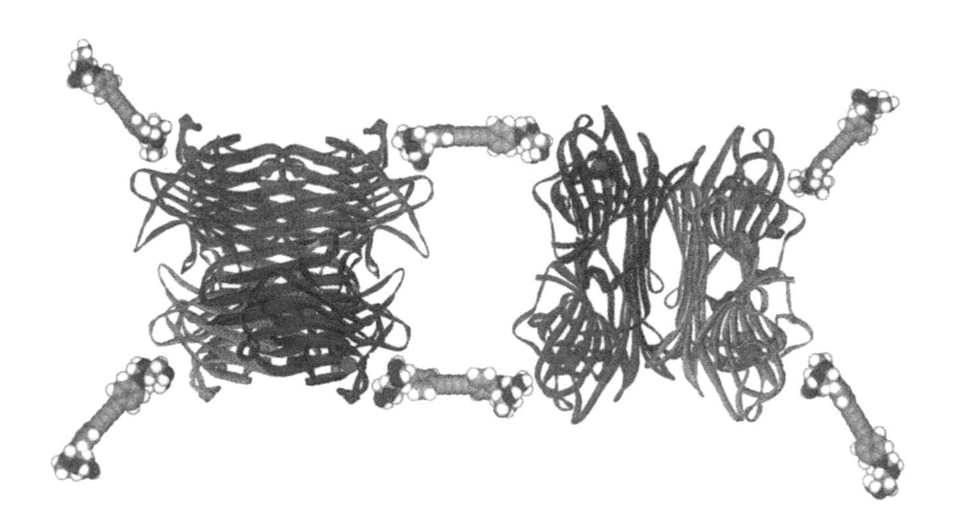

*Figure 8.2. Schematic representation of Concanavalin A lectin tetramer forming a cross-linked lattice with dimmer **18**.*

8. Turnbull, W. B.; Stoddart, J. F. *Rev. Molecul Biotechnol.* **2002,** *90*, 231-255.
9. Nagahori, N.; Lee, R. T.; Nishimura, S.-I.; Pagé, D.; Roy, R.; Lee, Y. C. *ChemBioChem* **2002,** *3*, 836-844.
10. Lee, Y. C.; Lee, R. T. *Acc. Chem. Res.* **1995,** *28*, 321-327.
11. Dam, T. K.; Roy, R.; Pagé, D.; Brewer, C. F. *Biochemistry* **2002,** *41*, 1351-1358.
12. Sharon, N. *FEBS Letters* **1987,** *217*, 145-157.
13. Arya, P.; Kutterer, K. M. K.; Qin, H.; Roby, J.; Barnes, M. L.; Kim, J.-M.; Roy, R. *Bioorg. Med. Chem. Lett.* **1998,** *8*, 1127-1132.
14. Vrasidas, I.; André, S.; Valentini, P.; Böck, C.; Lensch, M.; Kaltner, H.; Liskamp, R. M. J.; Gabius, Hans-J.; Pieters, R. J. *Org. Biomol. Chem.* **2003,** *1*, 803-810.
15. Sörme, P.; Qian, Y.; Nyholm, P.-G.; Leffler, H.; Nilsson, U. J. *ChemBioChem.* **2002,** *3*, 183-189.
16. André, S.; Liu, B.; Gabius, H.-J.; Roy, R. *Org. Biomol. Chem.* **2003,** *1*, 3909-3916.
17. Zaccai, N. R.; Maenaka, K.; Maenaka, T.; Crocker, P. R.; Brossmer, R.; Kelm, S.; Jones, E. Y. *Structure* **2003,** *11*, 557-567.
18. Stahl, P. D. *Curr. Opin. Immunol.* **1992,** *4*, 49-52.
19. Seddiki, N.; Rabehi, L.; Benjouad, A.; Saffar, L.; Ferriere, F.; Gluckman, J.-C.; Gattegno, L. *Glycobiology* **1997,** *7*, 1229-1236.
20. Pagé, D.; Roy, R. *Bioconjugate Chem.* **1997,** *8*, 714-723.
21. Pagé, D.; Zanini, D.; Roy, R. *Biorg. Med. Chem.* **1996,** *4*, 1949-1961.
22. Pagé, D.; Roy, R. *Glycoconjugate J.* **1997,** *14*, 345-356.
23. Sonogashira, K. In Diederich, F. and Stang, P. J., Eds.; *Metal-Catalyzed Cross-Coupling Reactions*; Wiley-VCH, Weinheim, Germany, 1998; pp 203-209.
24. Das, S. A.; Trono, M. C.; Roy, R. *Methods Enzymol.* **2003,** *362*, 3-17.
25. Mereyala, H. B.; Gurrala, S. R. *Carbohydr. Res.* **1998,** *307*, 351-354.
26. Roy, R.; Das, S. K.; Santoyo-González, F.; Hernández-Mateo, F.; Dam, T. K.; Brewer, C. F. *Chem. Eur. J.* **2000,** *6*, 1757-1762.
27. For a review: Siemsen, P.; Livingston, R. C.; Diederich, F. *Angew. Chem. Int. Ed.* **2000,** *39*, 2632-2657.
28. Hay, A. S. *J. Org. Chem.* **1962,** *27*, 3320-3321.
29. Kaufman, R. J.; Sidhu, R. S. *J. Org. Chem.* **1982,** *47*, 4941-4947.
30. Epstein, J.; Eichbaum, Q.; Sheriff, S.; Ezekowitz, R. A. B. *Curr. Opin. Immunol.* **1996,** *8*, 29-35.
31. Kawasaki, T. *Biochim. Biophys. Acta* **1999,** *1473*, 186-195.
32. Saito, S.; Yamamoto, Y. *Chem. Rev.* **2000,** *100*, 2901-2915.
33. Roy, R.; Das, S. K.; Dominique, R.; Trono, M. C.; Hernández-Meteo, F.; Santoyo-González, F. *Pure Appl. Chem.* **1999,** *71*, 565-571.

34. Dominique, R.; Liu, B.; Das, S. K.; Roy, R. *Synthesis* **2000,** *6*, 862-868.
35. Gan, Z.; Roy, R. *Tetrahedron Lett.* **2000,** *41*, 1155-1158.
36. Grubbs, R. H.; Chang, S. *Tetrahedron* **1998,** *54*, 4413-4450.
37. Fürstner, A. *Angew. Chem. Int. Ed.* **2000,** *39*, 3012-3043.
38. Das S. K.; Roy, R. *Tetrahedron Lett.* **1999,** *40*, 4015-4018.
39. Dam, T. K.; Brewer, C. F.; *Chem. Rev.* **2002**, *102*, 387-429.
40. Pagé, D.; Roy, R. *Int. J. Biochrom.* **1996**, *3*, 231-244.
41. Moothoo, D. N.; McMahon, S..A.; Dimick, S. M.; Toone, E. J.; Naismith, J. *Acta Crystal. D: Biol. Crystal.* **1998,** *D54*, 1023-1025.

Chapter 9

Antifreeze Glycoprotein Analogs: Synthesis, In Vitro Testing, and Applications

Vincent Bouvet and Robert N. Ben*

Department of Chemistry, D' Iorio Hall, University of Ottawa, Ottawa, Ontario, Canada K1N 6N5

A series of first generation C-linked antifreeze glycoprotein (AFGP) analogs have been successfully prepared using conventional solid phase chemistry. These glycoconjugates range in molecular weight between 1.5 to 4.1 Kda and can be prepared using traditional linear solid phase protocol. Unlike the native system, the C-linked analogs possess enhanced chemical and biological stability and consequently are well-suited for many potential medical, industrial and commercial applications. Despite dramatic structural modifications (relative to the native system), several of these first generation analogs display significant antifreeze protein-specific activity. This implies that the rational design and synthesis of low molecular weight AFGP possessing enhanced chemical and biological stability is a feasible and worthwhile goal.

Introduction

Biological antifreezes constitute a diverse class of proteins found in Arctic and Antarctic fish, as well as in amphibians, trees, plants and insects. These compounds are unique in that they have the ability to inhibit the growth of ice and consequently, are essential for the survival of organisms inhabiting environments where sub-zero temperatures are routinely encountered.

There are two types of biological antifreezes, the antifreeze proteins (AFPs) and antifreeze glycoproteins (AFGPs) (*1-3*). Antifreeze proteins are divided into four subtypes (Type 1-4) each possessing a very different primary, secondary and tertiary structure. In contrast, AFGPs are subject to considerably less structural variation. A typical AFGP is composed of a repeating tripeptide unit (threonyl-alanyl-alanyl) in which the secondary hydroxyl group of the threonine residue is glycosylated with the disaccharide β-D-galactosyl-(1,3)-α-D-N-acetylgalactosamine (Figure 1).

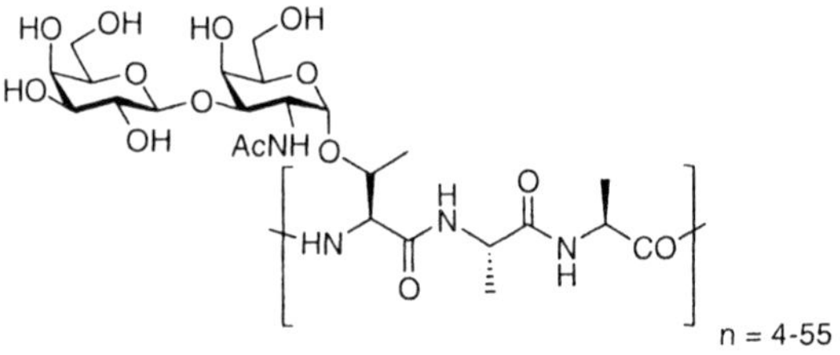

HO OH HO OH
HO O O O
OH AcNH O
HN H N O N H CO
O
n = 4-55

Figure 1. A typical antifreeze glycoprotein (AFGP).

Eight distinct AFGP subtypes exist; glycoproteins 20-33 Kda are referred to as AFGP 1-4 and those less than 20 KDa constitute AFGP 5-8. The lower molecular weight glycoproteins (AFGP 7-8) occasionally have the L-threonine residue substituted with L-arginine and one or both L-alanine residues substituted with L-proline (*4*).

The ability to inhibit the growth of ice has potential medical, industrial and commercial applications. Unfortunately, many of these applications have not been fully realized. One reason for this is that the isolation and purification of AFGP is a laborious and costly process often resulting in mixtures, making characterization difficult (*5*). Additional reasons include the fact that the AFGP mechanism of action is not understood at the molecular level and the nature of the protein-ice interface remains in question (*6*).

Mechanism of Action

Deep sea polar fish have evolved a mechanism to ensure their survival by preventing uncontrolled ice growth *in vivo*. They accomplish this by effectively lowering their freezing point below that the surrounding water. In the case of the polar oceans, this temperature is approximately -1.8 °C. Interestingly, this freezing point depression is not accomplished through the use of colligatively acting substances such as sugars or salts since, the delicate osmotic balance in the cell is not affected. Instead, these organisms employ unique proteins and glycoproteins to control the growth of ice.

Antifreeze glycoproteins have been shown to bind to the surface of ice and inhibit growth. During the last decade, there has been great interest in elucidating the mechanism by which this occurs but, despite these efforts the mechanism is not well understood. On a macroscopic level, the mechanism is regarded as an adsorption-inhibition process in which the biological antifreeze binds to the surface of a growing ice crystal *(7, 8)*. At this stage, growth occurs on ice surfaces between adjacent antifreeze molecules and these surfaces grow with a high surface curvature. Since the energetic cost of adding a water molecule to this convex surface is high, a non-equilibrium freezing point depression is observed while the melting point remains constant. This is referred to as the Kelvin Effect and the difference between melting and freezing points is defined as thermal hysteresis (TH). There are two models that rationalize ice growth inhibition based on a 2D or 3D Kelvin Effect; these are the mattress model and step pinning model. In the mattress model *(9)* (Figure 2), the adsorbed molecules prevent ice growth perpendicular to the ice surface, while in the step pinning model the molecules block the growth of a step *(8)*.

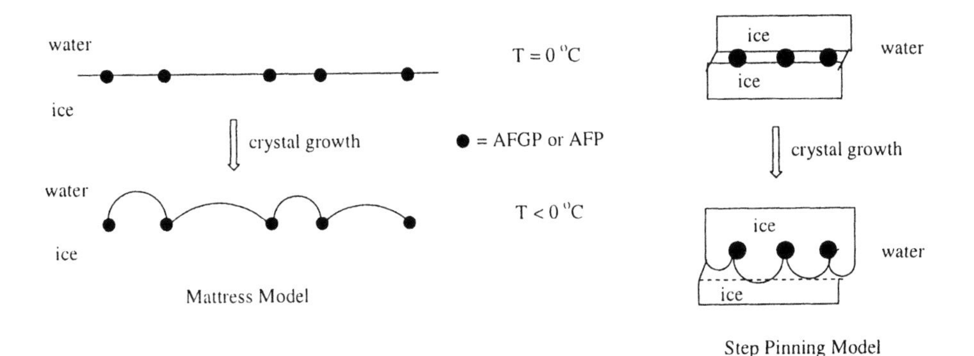

Figure 2. *Adsorption Inhibition Mechanisms*

Both models assume irreversible adsorption of AFGP onto the ice surface. However, neither model can explain the fact that high levels of adsorption are not observed at low concentrations. Consequently, alternate mechanisms have been proposed *(10)*.

At the molecular level, an understanding of how these molecules inhibit ice crystal growth remains a source of intense debate. Interactions between the antifreeze glycoprotein and ice surface are thought to be hydrogen bond based. However, modeling experiments have demonstrated that the number of hydrogen bonds formed between the antifreeze molecule and ice surface appears to be insufficient to explain the observed tight binding of the antifreeze to ice *(11)*. More recently, researchers have been divided over the importance of hydrogen bonding and its role in the mechanism of action. While it has been proposed that the hydrophilic interactions between polar residues and the water molecules on the ice surface are extremely important *(12)*, other researchers have invoked the idea that entropic and enthalpic contributions from hydrophobic residues are essential for adsorption onto the ice surface *(13)*. Despite the fact that significant entropic contributions are likely to be gained upon exclusion of water from the protein and ice surfaces, a detailed molecular mechanism invoking hydrophobic and/or hydrophilic interactions with emphasis on the role they play in adsorption of the antifreeze to the ice surface has failed to emerge.

In an effort to address this issue, the cooperative binding of antifreeze proteins as well as the role of side chain flexibility *(14, 15)* has been investigated. However, further complications have arisen with the discovery that different antifreeze proteins bind to separate faces or surfaces of an ice crystal *(2)*. It is not surprising then, that a unified hypothesis centered on the molecular mechanism of action has not been proposed.

Another obstacle to elucidating the molecular mechanism of action for AFGPs is that the ice-water interface has not been well characterized. This interface is not an abrupt transition as typically represented in the static models (Figure 2). In fact, recent evidence shows the loss of organized ice structure at the interface as being fairly gradual, occurring over approximately ten angstroms *(16)*. This is problematic when attempting to "map" possible interactions between the AFGP and ice surface. Since dynamic models of the ice-water interface have not been developed, static models continue to be used. Despite this limitation, valuable insight the mechanism of action has been obtained from structure-activity studies with AFPs and AFGPs since the essence of all structure-activity analyses originate from the receptor theory *(17)*.

Applications

Any organic compound with the ability to inhibit the growth of ice has many potential medical, commercial and industrial applications. Potential medical applications center on cryoadjuvants to protect cells from injury during cryopreservation or hyperthermic storage (*18-20*). In these instances, the most desirable compounds are those that are potent recrystallization-inhibitors and thus, prevent cellular damage during the thawing cycles or long-term storage (*21*). While there are many cryoadjuvants that are routinely used, application of these compounds is limited by their inherent toxicity.

Potent recrystallization-inhibitors will also be useful as additives to prevent recrystallization in materials such as grain oriented silicon steel or silicon-iron composites (*22, 23*). In a different venue, such compounds have applications in the food industry to enhance the texture and tastes of various frozen food products.

Unfortunately, commercialization of native AFGP for the above applications has yet to be realized. This is due to the fact that the bioavailability of native AFGP is very limited and the isolation/purification is a very laborious and costly process. Similarly, the chemical synthesis of AFGP is an equally costly and lengthy process despite the many recent advances in the synthesis or complex glycoconjugates. Finally, the chemical and biological instability of the anomeric carbon-oxygen glycosidic bond makes many *in vivo* and *in vitro* applications unrealistic. As a consequence, low molecular weight AFGP analogs with enhanced stability and efficacy that can be "custom-tailored" to individual applications are urgently required.

Our Approach

Despite the many advances in the field of oligosaccharide and glycoconjugate chemistry, complex glycans require lengthy and costly syntheses (*24*). Two reasons for this are the high lability of the anomeric carbon-oxygen bond under various reaction conditions and the need to employ orthogonal protecting group strategies. As part of our continuing efforts towards the rational design of functional AFGP analogs possessing enhanced chemical and biological stability, we have developed a general synthetic strategy to afford structural analogs of AFGP (*56*). Our approach is a building block one in which a core glycosylated tripeptide unit is assembled in a linear fashion using solid phase synthesis (Figure 3). One advantage of our methodology is that C-linked glycoconjugates are utilized. In other words, the carbon-oxygen anomeric bond is replaced with a carbon-carbon bond. We refer to the glycosylated tripeptide as a building block or monomer unit.

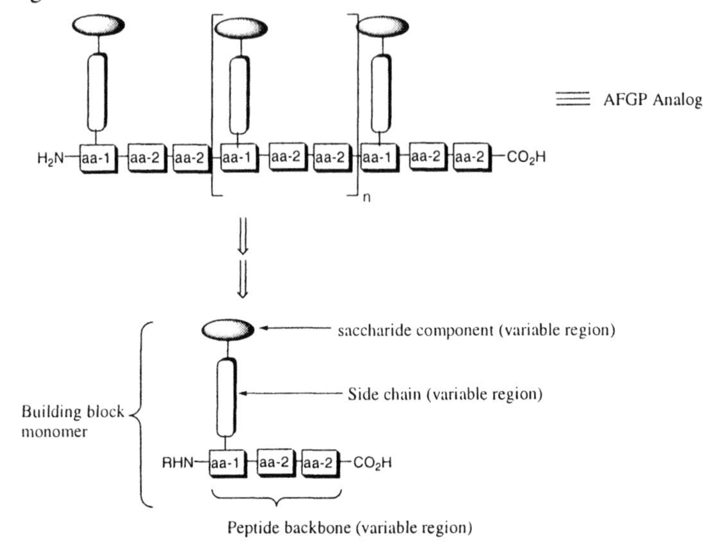

Figure 3. *AFGP Analogs.*

So far, all AFGP analogs prepared in our laboratory have utilized the L-lysine-glycine-glycine tripeptide unit that is radically different in structure than the core repeating tripeptide unit found in native AFGP. The threonine residue in native AFGP was replaced with lysine for two reasons. Firstly, recent work has demonstrated that an alanine-lysine rich polypeptide possessed weak antifreeze protein-specific activity (*26*).

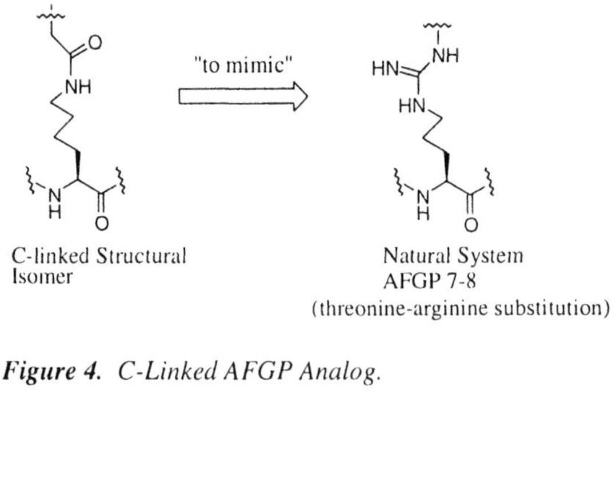

"to mimic"

C-linked Structural Isomer

Natural System
AFGP 7-8
(threonine-arginine substitution)

Figure 4. *C-Linked AFGP Analog.*

Secondly, the C-linked galactosyl-lysine residue in our analogs serves as a structural analog for the L-arginine residue occasionally substituted for L-threonine in AFGP 7-8 (Figure 4).

Synthesis of AFGP Analogs

The synthesis of the glycosylated building block or monomer unit is convergent in nature such that both the carbohydrate and tripeptide component are prepared separately and then coupled together prior to assembling the glycopolymer on solid phase support (*27*). Synthesis of the carbohydrate component is outlined in scheme 1. *β*-D-galactose was acetylated (acetic anhydride, pyridine, DMAP) in 96% isolated yield and was then converted to the galactosyl bromide (acetic acid, hydrogen bromide) **1** in 92% yield. The glycosyl bromide was subjected to a highly stereoselective photochemical-mediated C-allylation using allylphenyl sulfone in the presence of *bis*-(tributylstannane) to furnish the C-linked galactopyranose derivative (*28*).

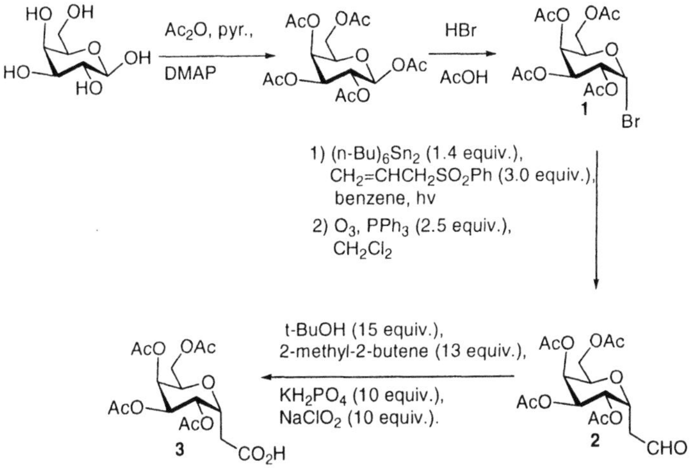

Ac_2O, pyr., DMAP

HBr, AcOH

1) $(n\text{-}Bu)_6Sn_2$ (1.4 equiv.), $CH_2{=}CHCH_2SO_2Ph$ (3.0 equiv.), benzene, hv
2) O_3, PPh_3 (2.5 equiv.), CH_2Cl_2

t-BuOH (15 equiv.), 2-methyl-2-butene (13 equiv.), KH_2PO_4 (10 equiv.), $NaClO_2$ (10 equiv.).

Scheme 1. *Synthesis of C-linked Carbohydrate Component.*

Ozonoylsis followed by reductive workup with triphenyl phosphine produced aldehyde **2** which was subsequently oxidized to furnish the C-linked galactosyl pyranose derivative **3**.

Preparation of the tripeptide component is outlined in Scheme 2. The glycine benzyl ester was coupled to commercially available *t*-Boc-glycine using 1,1-carbonyldiimidazole as coupling agent. Removal of the carbamate on the N-terminus (TFA) furnished **4** in 90% yield (two steps). Compound **4** was coupled to an orthogonally protected lysine derivative to produce the tripeptide **5**. After removal of the carbamate protecting group on the ε-amino terminus, **6** was coupled with the C-linked pyranose derivative to furnish **7** in 70% yield. Hydrogenolysis produced the requisite building block for solid phase synthesis.

Typical solid phase synthesis protocols were employed to assemble the tripeptide building blocks into C-link AFGP glycopolymers. The method is illustrated in Scheme 3.

Glycoconjugates were prepared on commercially available Wang resin pre-loaded with Fmoc-glycine (Novabiochem). Loading of the resin was verified prior to the preparation of each glycopolymer. Coupling of each tripeptide building block was accomplished via iterative treatment with a 20% piperidine/DMF solution followed by building block (3.2 equiv.), *O*-(7-azabenzotriazol-1-yl)-N,N,N',N'-tetramethyluronium hexafluorophosphate (HATU, 3.0 equiv.) and diisopropylamine (3.0 equiv.) in DMF. The glycopeptide was removed from the resin bead by treatment with 50% TFA and the acetate protecting groups removed by treatment with sodium methoxide solution. Glycopeptides **9**, **10** and **11** were produced in 43-90% chemical yield after purification by reversed phase HPLC.

In Vitro Testing of AFGP Analogs

The first generation C-linked analogs were assayed using two conventional techniques. These techniques are recrystallization-inhibition (RI) assay *(29, 30)* and nanoliter osmometry *(31)*. The RI assay offers a distinct advantage in that it is effective at detecting antifreeze protein-specific activity at concentrations typically too low to cause a thermal hysteretic effect *(29)*.

Graph 1 shows the RI assay results with our first C-linked AFGP analogs. The monomer unit as well as **9** (3-mer), **10** (6-mer) and **11** (9-mer) were assayed. All measurements were performed in triplicate. The Y-axis depicts the mean largest grain size (MLGS in mm^2) of ice crystals measured directly from photographs of the sample as described previously *(29)*. To circumvent false positives due to additive carbohydrate concentrations amongst samples, the carbohydrate concentration of each sample was corrected to 0.021 mmol/L. For instance, compound **9** is 0.007 mmol/L with respect to protein concentration but is 0.021 mmol/L with respect to carbohydrate concentration. Similarly, compound **10** is 0.021 mmol/L with respect to carbohydrate but 0.0033 mmol/L with respect to peptide. Both compound **11** and commercially available glycosylated Bovine Serum Albumin (BSA-conj., purchased as 20 mmol of β-D-galactose/mmol of peptide) were treated in a similar fashion. The peptide control ((L-lysine-glycine-glycine)$_6$-glycine) is 0.021 mmol/L. PBS is used as the control since all samples are tested in a PBS solution.

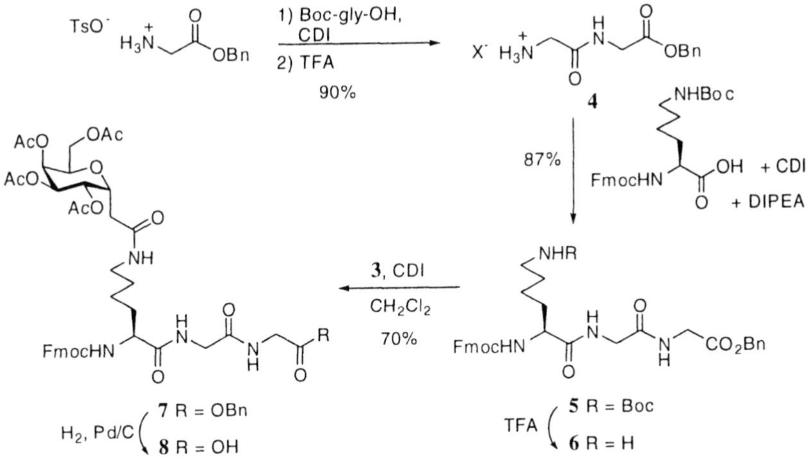

Scheme 2. Synthesis of Glycosylated Tripeptide Component.

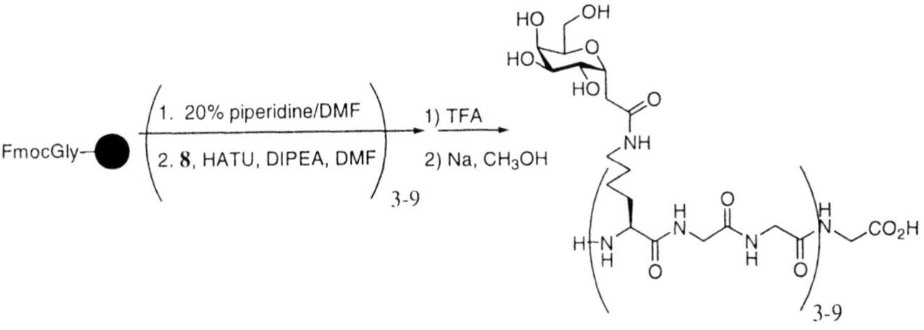

Scheme 3. Solid Phase Synthesis of C-linked AFGP Analog.

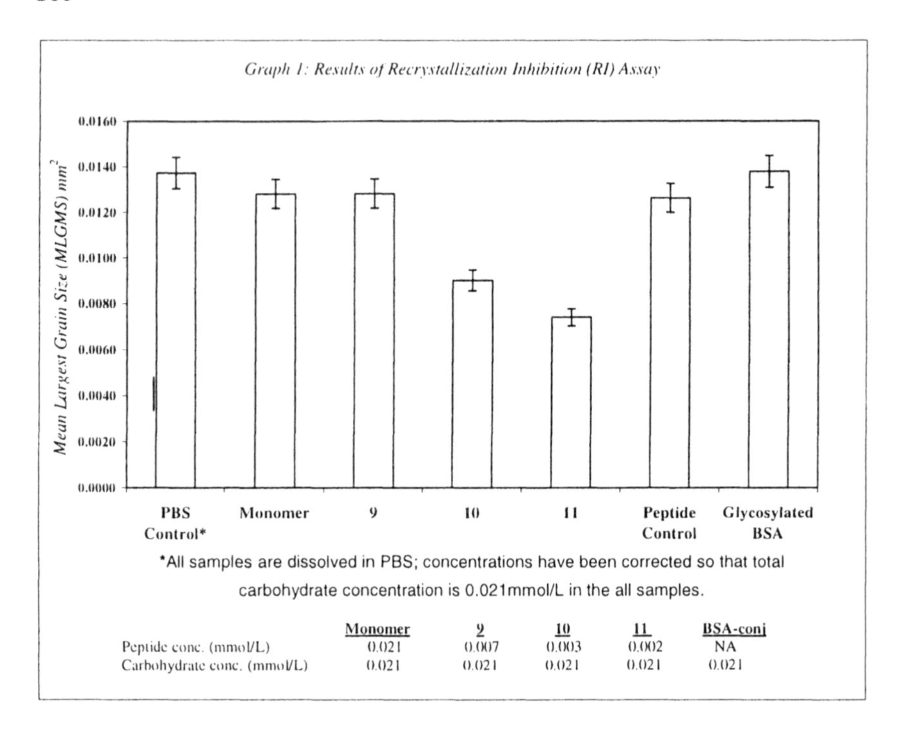

*All samples are dissolved in PBS; concentrations have been corrected so that total carbohydrate concentration is 0.021mmol/L in the all samples.

	Monomer	9	10	11	BSA-conj
Peptide conc. (mmol/L)	0.021	0.007	0.003	0.002	NA
Carbohydrate conc. (mmol/L)	0.021	0.021	0.021	0.021	0.021

The building block and compound **9** do not show any RI activity relative to PBS. However, compounds **10** and **11** possess RI activity. This is remarkable given the structural modifications of these compounds relative to AFGP 8. Furthermore, it is an interesting observation that **11** (n = 9) appears to be slightly more active than **10** (n = 6). This trend is consistent with the observation that lower molecular weight AFGPs (fractions 5-8) are less active than AFGP 1 *(32)*. Mindful of the relationship between glycoprotein length and antifreeze protein-specific activity **3** and **4** were tested against an authentic sample of native AFGP-8, generously donated by AF Protein Inc. (data not shown on graph). AFGP-8 is the smallest of the AFGPs (n = 4, 2.2 Kda) and is approximately 20X less active than AFGP 2-5. A direct comparison revealed that our analogues are weakly active (i.e. AFGP-8 is approximately ninety times more active than compound **10**).

Non-specific RI effects (i.e. inhibition of ice growth) are common with colligatively acting substances such as inorganic salts, glycerol and oligosaccharides. In order to confirm that the activity of **3** was not a non-specific RI effect, several controls were also tested. The first was a peptide control ((L-Lysine-glycine-glycine)$_6$-glycine) composed of six tripeptide units analogous to **3**, but with no sugars attached to the lysine side chains. As expected, this sample did not inhibit the growth of ice at concentrations equal to or even twice that of **3** and **4**, highlighting the importance of the carbohydrate residues. This observation is consistent with earlier work demonstrating that the disaccharide residues in native AFGP are crucial to activity *(33)*. It was not possible to test the peptide control in the presence of uncoupled galactose because non-specific RI effects would be produced. Commercially available glycosylated BSA was also tested at a concentration of 0.21 mmol/L (relative to carbohydrate). As illustrated in Graph 1, no RI activity was detected, suggesting that glycosylation is not the only important factor for RI activity. This result confirms that the antifreeze protein-specific activity observed with **10** and **11** is genuine and not the result of a non-specific RI effect.

AFGP analogs **9**, **10** and **11** were also tested for TH activity using nanoliter osmometry at concentrations identical to those used in the RI assay. These measurements confirm that **10** and **11** induce a small thermal hysteretic gap of 0.056 °C (30 mosmol). Initially we were bothered by the fact that **10** and **11** appeared to display similar activity in the TH assays and different activity in the RI assay. However, recent results have shown that the relationship between RI and TH activity is qualitative and not quantitative *(34)*.

In addition to the observed TH gap, both **10** and **11** possess the ability to bind to ice as evidenced by unusual ice crystal morphology in the nanoliter osmometry assay. This "dynamic ice shaping" ability is a property unique to biological antifreezes *(31)* and occurs when a biological antifreeze binds to the surface of an ice crystal. Figure 5 (a) depicts a single ice crystal in the absence of biological antifreeze. Notice that the crystal is perfectly round and has no facets or edges. Identical images have been obtained when a single crystal is grown in the presence of BSA, glycosylated-BSA and sodium chloride.

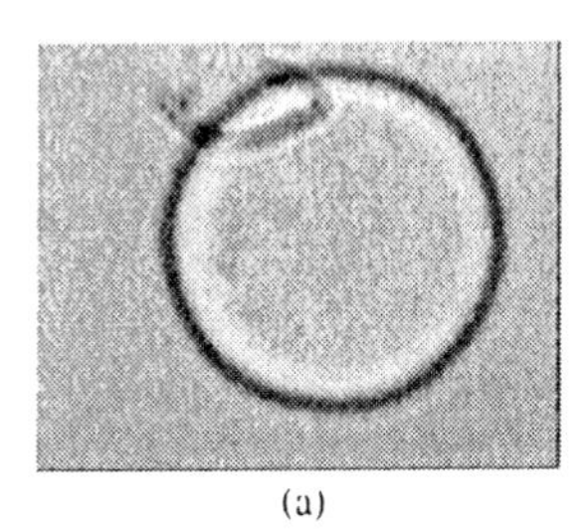

(a)

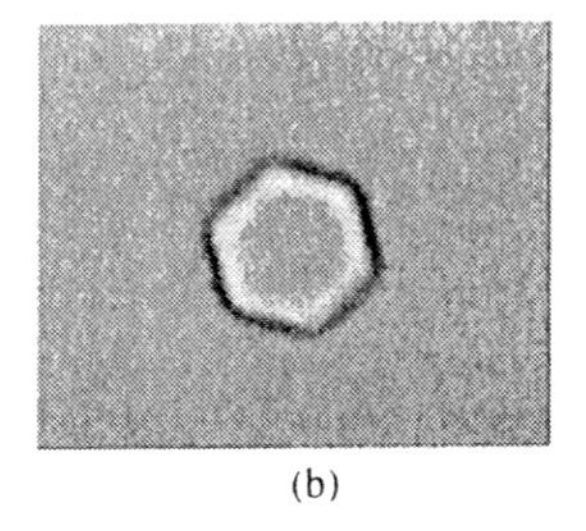

(b)

Figure 5. *Images of Single Ice Crystal Using Nanoliter Osmometry. (a) doubly distilled water, (b) 0.0033 mmol/L solution of* **10** *in doubly distilled* H_2O.

Figure 5 (b) depicts a 0.0033 mmol/L solution (peptide concentration) of **10** in doubly distilled water. The single crystal is hexagonal, indicative of dynamic ice shaping and this hexagonal shape has been previously reported with weakly active mutants of the Type I AFP *(35)*. Identical images were obtained for compound **11**. These results verify that C-linked AFGP analogues **10** and **11** possess weak antifreeze protein-specific activity.

Conclusions

Several first generation C-linked AFGP analogs ranging in molecular weight from approximately 1.5 to 4.1 Kda have been synthesized. Structurally, these analogs are very different from native AFGP 8. However, despite these differences, two of the analogs (compounds **10** and **11**) display significant RI activity and induce very small TH gaps. In addition, both compounds possess the ability to bind to ice, as evidenced by dynamic ice shaping, a trait unique to biological antifreezes. Collectively, these results verify that compounds **10** and **11** are functioning in a manner similar to that of native AFGP. This is the only example where an artificial C-linked AFGP analog possessing enhanced stability also demonstrates antifreeze protein-specific activity. The fact that the stability of the glycosidic linkage between the carbohydrate and peptide backbone is greatly enhanced means that these "artificial antifreezes" are more suitable for many medical, commercial and industrial applications. More importantly, these results suggest that the rational design and synthesis of chemically and biologically stable artificial antifreeze glycoproteins is a reasonable and worthy goal.

Experimental

L-Lys-[2-(α-D-galactopyranosyl)acetamide]-Gly-Gly-OH (building block):

The synthesis of the L-Lys-[α-D-galactopyranosyl)acetamide]-Gly-Gly-OH was reported previously *(27)*. Fmoc-L-Lys-[2-(2,3,4,6-tetra-*O*-acetyl-α-D-galactopyranosyl)acetamide]-Gly-Gly-OH (100 mg, 0.12 mmol) was dissolved in methanol (10 mL) and solid sodium methoxide (30 mg, 0.5 mmol) was added. The solution was stirred overnight at room temperature. After this time, acidic ion-exchange resin (IR 120) was added to neutralize the basic solution and the solution was then filtered and concentrated. The residue was re-dissolved in a 1:1 mixture of methanol/water and then washed with hexanes. The methanol/water layer was concentrated under vacuum to afford 40 mg (75% yield) of **1** as a colorless oil. ^{1}H NMR (360 MHz, D_2O) δ: 4.5 (m, 1H), 4.0 (m, 5H), 3.84 (m, 3H), 3.76 (dd, J = 9.9, 3.3 Hz, 1H), 3.71 (d, J = 6.0 Hz, 2H), 3.23 (t, J = 6.7 Hz, 2H), 2.69 (dd, J = 14.9, 10.7 Hz, 1H), 2.60 (dd, J = 14.9, 4.4 Hz, 1H), 1.90 (m, 2H), 1.58 (m, 2H) 1.47 (m, 2H); ^{13}C NMR (90 MHz, D_2O) δ:

178.7, 175.6, 173.0, 172.0, 75.4, 74.9, 72.1, 71.2, 70.0, 63.3, 55.6, 44.7, 41.3, 34.8, 32.8, 30.3, 23.9; LRMS (electrospray, H_2O, positive ion mode) calcd for $C_{18}H_{32}N_4O_{10}$ (M^+) 464.2; found 487.1 (M^+ + Na)

***Tris*-[*L*-Lys-[2-(α-D-galactopyranosyl)acetamide]-Gly-Gly]-Gly-OH (9):**

The solid phase synthesis was performed on Fmoc-Gly-Wang resin from Nova Biochem. Fmoc-Gly-Wang resin (100 mg, 0.075 mmol active site) was swollen in DMF for 1hr and then treated with 20% piperidine in DMF for 30 min to remove the Fmoc protecting group. The flask was then charged with DMF (5 mL), Fmoc-L-Lys-[2-(2,3,4,6-tetra-*O*-acetyl-α-D-galacto-pyranosyl)acetyl-Gly-Gly-OH (193 mg, 0.225 mmol), N,N-Diisopropylethylamine (0.04 mL, 0.225 mmol) and HATU (86 mg, 0.225 mmol). This solution was allowed to stir until negative Kaiser and TNBS tests were obtained. The reaction sequence described above was repeated three times. After the third coupling was complete, a 20% piperidine in DMF solution was added and the solution stirred for 30 mins. The resin was then washed successively with DMF, 4% acetic acid, dichloromethane and methanol and dried over KOH under vacuum for 14 hrs. Cleavage of the glycopeptide from the resin was accomplished by treatment with a 1:1 mixture of trifluoroacetic acid and dichloromethane for 30 minutes. The solution was filtered off and the resin washed with trifluoroacetic acid. Diethyl ether was then added to the filtrate and a white precipitate resulted. The white precipitate was filtered and dissolved in methanol (5 mL) with a catalytic amount of sodium metal. After stirring the reaction overnight the solvent was removed in vacuo to furnish a yellow residue. Purification by reversed-phase HPLC followed by lyophilization afforded 38 mg (90% yield) of **9**. IR (KBr disk) 3302, 1743, 1656 cm^{-1}; ^1H NMR (360 MHz, D_2O) δ:4.60 (3H, br. s), 4.05 (17 H, m), 3.85 (3 H, br. d, J = 5.9 Hz), 3.78 (9 H, m), 3.72 (6 H, d, J = 5.2 Hz), 3.23 (6 H, br. s), 2.70 (3 H, dd, J = 2.0, 11.2 Hz), 2.60 (3 H, d, J = 9.8 Hz), 1.90 (6 H, br. s), 1.56 (6 H, br. s), 1.36 (6 H, br. s) ; ^{13}C NMR (90 MHz, D_2O) δ:178.4, 178.1, 175.2, 172.1, 77.8, 77.3, 74.5, 73.6, 72.5, 65.7, 58.7, 53.7, 47.5, 47.2, 54.9, 44.0, 53.8, 37.2, 35.2, 32.7, 27.3, 24.9; MS (electrospray, H_2O, negative ion mode) calcd for $C_{56}H_{95}N_{13}O_{29}$ (M^+) 1413.64; found 1452.50 (M^+ + K).

[L-Lys-[2-(α-D-galactopyranosyl)acetamide]-Gly-Gly]$_6$-Gly-OH (10):

A protocol analogous to the preparation of **9** was followed to afford 38 mg (43 % yield) of **10** as a white solid after purification via reversed-phase HPLC. IR (KBr disk) 3310, 1741, 1652 cm^{-1}; ^1H NMR (360 MHz, D_2O), δ: 4.47 (6 H, br. s), 3.97 (36 H, m), 3.81 (6 H, br. s), 3.74 (6 H, dd, J = 9.6 Hz, 1.1 Hz), 3.67 (20 H, m), 3.13 (24 H, t, J = 7.0 Hz), 2.69-2.58 (12 H, m), 2.05 (8 H, s), 1.74 (24 H, m), 1.61 (16 H, m); ^{13}C NMR (90 MHz, D_2O) δ: 174.3, 172.9, 171.5, 169.9, 71.1, 70.5, 67.8, 66.9, 65.7, 59.6, 58.9, 52.1, 42.7, 40.7, 40.4, 39.4,

37.3, 30.5, 28.4, 25.9, 20.5, 20.3, 19.7, 19.6; MS (MALDI) calcd for $C_{110}H_{185}O_{56}N_{25}$ (M^+) 2753.4955, found 2776.7990 (M^+ + Na).

[L-Lys-[2-(α-D-galactopyranosyl)acetamide]-Gly-Gly]$_9$-Gly-OH (11):
A protocol analogous to the preparation of **2** was followed to afford 62 mg (41 % yield) of **11** as a colorless oil after purification via reversed-phase HPLC. IR (KBr disk) 3317, 1732, 1668 cm^{-1}; ^{1}H NMR (360 MHz, D_2O) δ: 4.49 (9 H, br. s), 4.08-3.88 (63 H, m), 3.80 (9 H, t, J = 6.4 Hz), 3.74 (9 H, dd, J = 9.4 Hz, 1.08 Hz), 3.66 (26 H, d, J = 7.6 Hz), 3.17 (30 H, m), 2.68-2.51 (18 H, m), 2.04 (12 H, d, J = 4.5 Hz), 1.79 (24 H, br. s), 1.49 (36 H, br. s); ^{13}C NMR (90.5 MHz, D_2O) δ 172.9, 171.5, 170.1, 169.0, 71.1, 70.6, 67.6, 66.9, 65.7, 56.9, 52.2, 52.1, 40.70, 40.4, 39.2, 39.1, 37.3, 30.5, 26.5, 26.0, 20.6, 19.6; MS (electrospray, H_2O, negative ion) calcd for $C_{164}H_{275}O_{83}N_{37}$ (M^+) 4090.8435, found 4090.8451.

Recrystallization-Inhibition Assay

The typical procedure for this technique is as follows. A 10 μL sample (as a PBS solution) is dropped down a three meter tube onto a polished aluminum surface (pre-cooled to – 80 °C) sitting on a block of dry ice. The sample is frozen instantaneously and transferred using razor blade to a refrigerated microscope stage preset at a -6 °C annealing temperature. The ice wafers are then photographed after thirty minutes at 30X magnification through crossed polarizing filters using a Nikon CoolPix 5000 digital camera. The mean largest grain size is then calculated and plotted relative to phosphate buffered saline (PBS) solution (*30*). This assay is an attractive alternative to other RI assays in that the degree of RI activity is quantifiable. In the absence of biological antifreeze, a thirty minute period is insufficient to observe a recrystallization-inhibition effect and thus, an increase in size and a decrease in the number of ice crystals is observed. In the presence of a biological antifreeze (which has the ability to inhibit the recrystallization process), much smaller crystals and an increased number of crystals is observed.

Acknowledgements

RNB holds a Canada Research Chair (Tier 2) in Medicinal Chemistry and acknowledges the National Institutes of Health (GM 60319), Petroleum Research Fund (ACS-PRF# 35280-G1) and AF Protein Inc. for financial support.

References

1. Yeh, Y.; Feeney, R. E. *Chem. Rev.* **1996**, *96*, 601-617.
2. Davies, P. L.; Sykes, B. D. *Curr. Opin. Struct. Biol.* **1997**, *7*, 828-834.
3. Fletcher G. L., Hew C. L., Davies P. L. *Annu. Rev. Physiol.* **2001**, *63*, 359-390.
4. Hew, C. L.; Slaughter, D.; Fletcher, G.; Shashikant, J.B. *Can. J. Zool.* **1981**, *59*, 2186.
5. Jiaang, J. W.; Hsiao, K. F.; Chen, S. T.; Wang, K. T. *Synthesis* **1999**, *9*, 1687-1690.
6. Karim, O. A.; Haymet, A. D. J. *J. Chem. Phys.* **1988**, *89*, 6889-6896.
7. Brown R. A.; Feeney R. E. *Biopolymers* **1985,** *24*, 1265-1270.
8. Wilson P. *Cryo-Lett.* **1993,** *14*, 31-36.
9. Knight C. A.; Cheng C. C.; Devries A. L. *Biophys. J.* **1991,** *59*, 409-418.
10. Hall D. G.; Lips A. *Langmuir* **1999,** *15*, 1905-1912.
11. Knight C. A.; Driggers E.; Devries A. L. *Biophys. J.* **1993,** ***64***, 252-259.
12. Wierzbicki A.; Taylor M. S.; Knight C. A.; Madura J. D.; Harrington J. P.; Sikes C. S. *Biophys. J.* **1996,** *71*, 8-18.
13. Chao H. M.; Houston M. E. Jr.; Hodges R. S.; Kay C. M., Sykes B. D.; Loewen M. C.; Davies P. L.; Sonnichsen F. D. *Biochemistry* **1997,** *36*, 14652-14660.
14. DeLuca C. I.; Comley R.; Davies P. L. *Biophys. J.* **1998,** *74*, 1502-1508.
15. Gronwald W.; Chao H.; Reddy D. V.; Davies P. L.; Sykes B. D.; Sonnichsen F. D. *Biochemistry* **1996,** *35*, 16698-16704.
16. Karim O. A.; Haymet A. D. J. *J. Chem. Phys.* **1988,** *89*, 6889-6896.
17. *Quantitative drug design: a critical introduction.* Martin Y. C., Marcel Decker, Inc., New York, NY, **1978**.
18. Hays, L. M.; Feeney, R. E.; Crowe, L. M.; Crowe, J. E.; Oliver, A. E. *Proc. Natl. Acad. Sci. USA* **1996**, *93*, 6835-6840.
19. Tablin, F.; Oliver, A. E.; Walker, N. J.; Crowe, L. M.; Crowe, J. H. *J. Cell. Phys.* **1996**, *165*, 305-313.
20. Hansen, T. N.; Smith, K. M.; Brockbank, K. G. M. *Transplantation Proceedings* **1993**, *25*, 3182-3184.
21. Knight, C. A.: Wen, D.; Laursen, R. A. *Cryobiology* **1995,** *32,* 23-34.
22. Harase, J.; Shimizu, R. *Nippon Kinzoku Gakkaishi* **1990,** *54,* 1-8.
23. Abbruzzese, G.; Ciancaglioni, I.; Campopiano, A. *Textures and Microstructures* **1988,** *8-9,* 401-412.
24. Lowary, T.; Meldal, M.; Helmboldt, A.; Vasella, A.; Bock, K. *J. Org. Chem.* **1998**, *63*, 9668.
25. Eniade, A.; Murphy, A. V.; Landreau, G.; Ben, R. N. *Bioconjugate Chemistry* **2001,** *12,* 817-823.
26. Wierbicki, A.; Knight, C.A.; Rutland, T.J.; Muccio, D.D.; Pybus, B.S.; Sikes, C.S. *Biomacromolecules* **2000**, *1*, 274.

27. Ben, R. N.; Eniade, A.; Hauer, L. *Organic Lett.* **1999,** *11,* 1759-1762.
28. Ponten, F. ; Magnusson, G. *J. Org. Chem.* **1996,** *61,* 7463-7466.
29. Horwath, K. L.; Easton, C. M.; Poggioli, G. J., Jr.; Myers, K.; Schnorr, I. L. *Eur. J. Entomol.* **1996,** *93,* 419-433.
30. Enaide, A.; Purushotham, M.; Ben, R. N.; Wang, J. B.; Horwath, K. *Cell Biochemistry and Biophysics* **2003,** *38,* 115-124.
31. Houston, M. E., Chao, H., Hodges, R. S., Sykes, B. D., Kay, C. M., Sonnichsen, F. D., Loewen, M. C., Davies, P. L. *J. Biol. Chem.* **1998,** *273,* 11714-11718.
32. Ananthanarayanan, V. S. *Life Chemistry Reports* **1989,** *7,* 1-32.
33. Feeney, R. E.; Yeh, Y. *Adv. Protein Chem.* **1978,** *32,* 191-282.
34. Sidebottom, C.; Buckley, S.; Pudney, P.; Twigg, S.; Jarman, C.; Holt, C.; Telford, J.; McArthur, A.; Worrall, D.; Hubbard, R.; Lillford, P. *Nature* **2000,** 406, 256.
35. Chao, H.; Houston, M. E., Jr.; Hodges, R. S.; Kay, C. M.; Sykes, B. D.; Loewen, M. C.; Davies, P. L.; Sonnichsen, F. D. *Biochemistry* **1997,** *36,* 14652.

Chapter 10

Unprotected Oligosaccharides as Phase Tags for Solution-Phase Glycopeptide Synthesis with Solid-Phase Workup

Zhongwu Guo

Department of Chemistry, Case Western Reserve University, 10900 Euclid Avenue, Cleveland, OH 4416

A new method was developed for glycopeptide synthesis with unprotected glycosyl amino acids as key building blocks and "phase tags". After oligosaccharides were prepared and linked to appropriate amino acids, the carbohydrate moieties were deprotected. Glycopeptides were assembled upon the resultant glycosyl amino acids containing unprotected oligosaccharides. While all reactions for glycopeptide assembly were performed in homogeneous *N*-methylpyrrolidinone (NMP) solutions, the product of each reaction was expediently isolated from the reaction mixture by precipitation method via addition of a less polar solvent, such as diethyl ether. Therefore, the unprotected oligosaccharides were acting as phase tags to facilitate product isolation. The final synthetic targets were purified by HPLC. This method, which is defined as solution-phase glycopeptide synthesis with solid-phase workup, proved to be very efficient and expedient, and it has a number of advantages. It has thus been successfully employed to synthesize several complex glycopeptides, including one of CD52 with an especially acid-labile oligosaccharide chain.

Introduction

Glycoproteins play a central role in numerous biological and pathological events (*1,2*). However, it is very difficult, if not impossible, to get homogeneous glycoproteins from nature due to the microheterogeneity problem (*3*). A solution to the issue largely relies on chemical synthesis. To meet the growing demand for homogeneous glycoproteins or glycopeptides, many synthetic methods have been developed for them in the past decades. These methods include solution- and solid-phase syntheses using properly protected glycosyl amino acids as the key building blocks (*4-8*), as well as other elegant synthetic designs, such as chemoenzymatic synthesis utilizing enzymatic transglycosylations or enzymatic elongation of the peptide and carbohydrate chains (*9-15*) and chemoselective ligations between carbohydrates and peptides (*16-19*). Another very interesting but inadequately addressed method is solid-phase synthesis with free glycosyl amino acids as building blocks (*20-22*). An advantage of the last method is that it does not need final stage carbohydrate deprotection. Nevertheless, owing to the remarkable diversity and complexity of glycopeptide structures, it is difficult for any specific method to fulfill various kinds of synthetic endeavors, even though the established methods have been successfully utilized to synthesize numerous glycopeptides. Therefore, alternative methods are desirable.

Among the synthetic methods investigated so far, solid-phase synthesis is one of the most attractive (*5-7,20,23*), for solid-phase synthesis is effective and the product separation is simple. Its synthetic protocol is to assemble glycopeptides on a polymer support, and following completion of the assembly, glycopeptides are retrieved from the polymer support and finally carbohydrate moieties are deprotected. However, there are two potential problems. One is associated with retrieval of glycopeptides from the polymer support, which traditionally needs strong acid treatment, e.g. using 95% trifluoroacetic acid (TFA). This condition can affect acid-labile glycosidic linkages, such as those of fucose and sialic acid (*24*). Another problem is related to the final stage carbohydrate deprotection that can potentially affect the peptides (*25*).

Trying to create a useful method for glycopeptide synthesis by avoiding the potential problems mentioned above, we recently explored a new strategy using unprotected glycosyl amino acids as building blocks and phase tags (*26-28*), as shown in Figure 1. *First*, protected oligosaccharides are prepared and linked to the appropriate amino acids, and then the carbohydrate chains are deprotected. The resultant glycosyl amino acids with unprotected oligosaccharides are almost insoluble in most organic solvents except ones that are very polar. *Next*, the glycopeptide is elongated from one of the glycosyl amino acids. So, after a glycosyl amino acid is dissolved in a polar organic solvent, e.g. NMP, to form a homogeneous solution, other substrates and/or reagents are added for reaction. Amino acid or peptide deptotection and coupling can thus be carried out under homogeneous conditions. After the reaction is finished, to the reaction mixture is added a less polar solvent, such as diethyl ether, and the glycopeptide product will precipitate, but the excessive reagents and side products will remain in solution. Finally, the glycopeptide is separated by filtration and washed with the less polar organic solvent. *Thereafter*, the glycopeptide is dissolved in the polar organic solvent again for further elongation. The cycle of dissolution-reaction-precipitation-filtration-washing can be repeated until the target glycopeptide is obtained, which is eventually purified by HPLC.

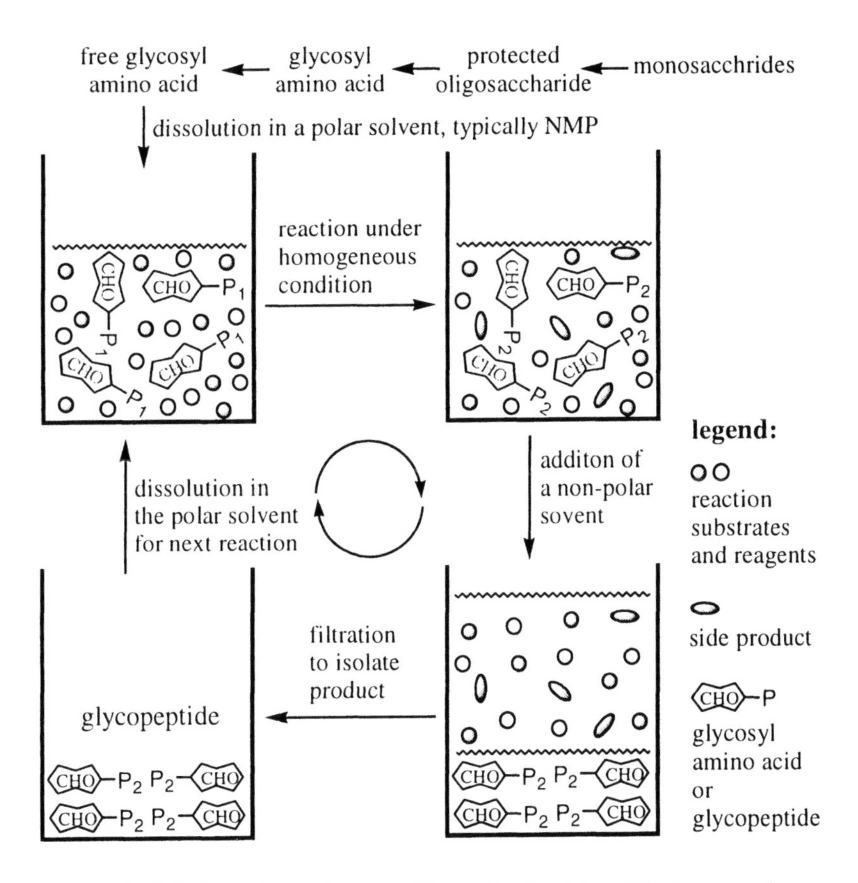

Figure 1. Solution-phase glycopeptide synthesis with solid-phase workup

In this synthetic scheme, the reactions for peptide elongation are carried out in homogeneous solutions, while the product of each reaction is isolated and briefly purified in solid form with free oligosaccharides serving as phase tags. It is therefore defined as "solution-phase glycopeptide synthesis with solid-phase workup". It can take advantage of both solution-phase and solid-phase syntheses, namely high reaction efficiencies of the former and simple product isolation of the latter. More importantly, the new synthetic method can avoid the problems associated with traditional solid-phase synthesis. For instance, even though free carbohydrate chains are used as phase tags, they are also a part of the synthetic targets and they need not to be removed after the synthesis is accomplished. Meanwhile, since the synthesis uses unprotected oligosaccharide substrates, final stage carbohydrate deprotection is not needed. An additional advantage of this method is that the reactions can be monitored by conventional methods, e.g. TLC, NMR and MS, as no polymers are involved.

Results and Discussion

The new synthetic method was first tested in the synthesis of some simple *N*-linked glycopeptides **1**, **2** and **3** (*26,28*). These targets were assembled from corresponding glycosyl asparagine via elongating the peptide chains along their *N*-termini by conventional Fmoc chemistry (*26,28*). NMP was used to dissolve substrates for the reactions, while diethyl ether was used to precipitate products. Thus, cleavage of Fmoc was realized in 25% piperidine in NMP, and coupling reactions were performed in anhydrous NMP. All intermediates and products were isolated by addition of 5-10 equivalents of diethyl ether to the reaction mixture. The target glycopeptides were obtained in excellent overall yields (over 80%). It was also noticed that oligosaccharides are more effective phase tags than monosaccharides.

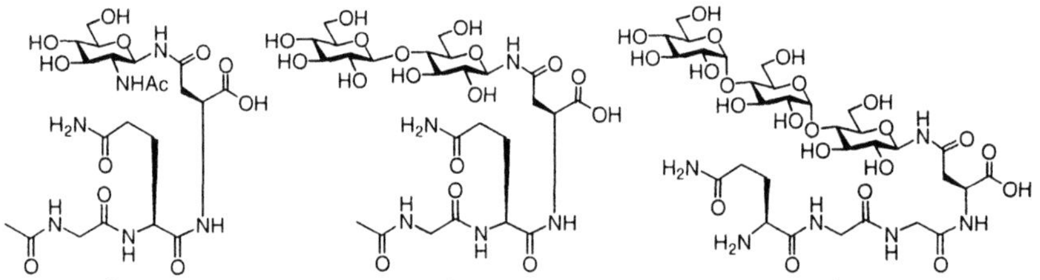

The method was also applied to **4**, which requires elongation of the peptide chain at both termini (Scheme 1). Thus, after the *N*-terminal elongation to get **5**, its *C*-terminus was deprotected with 50% TFA in dichloromethane (DCM). The free carboxyl group was thereby activated by selective reaction with 1-hydroxybenzotriazole (HOBt) and *N,N*-dicyclohexylcarbodiimide (DCC). The resultant active ester **7** was coupled to the benzyl ester of glycine to give the expected glycopeptide **4** that was finally isolated in an overall yield of 85% (*26*).

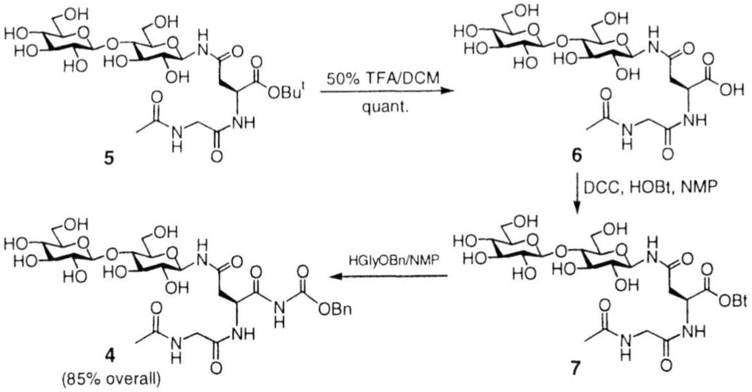

Scheme 1. Glycopeptide assembly via C-terminal elongation

Even though the acidic treatment utilized to deprotect *C*-terminus was not a problem in the synthesis of **4**, it is indeed a potential concern in some cases. For example, if *C*-terminal peptide sequence were longer, repeated acidic treatments would become necessary, which might affect the carbohydrate moieties. To avoid this complex situation, we conceived a more convergent design shown in Scheme 2 (*28*). After the unprotected glycosyl amino acid is obtained, its peptide chain is elongated along the *N*-terminus. Then, its *C*-terminus is deprotected and activated to give **8**. On the other hand, the *C*-terminal peptide **9** can be prepared by conventional methods. Finally, the two fragments are coupled to afford the target structure **10**. This design can minimize the number of steps of *C*-terminal deprotection and elongation of glycopeptides.

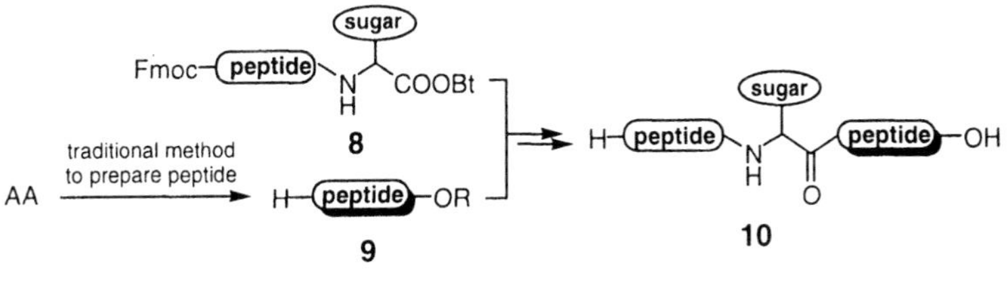

Scheme 2. Convergent assembly of glycopeptides via fragment coupling

Fragment coupling method was first applied to glycopeptide **11** (Scheme 3). The *C*-terminal peptide **12** was prepared by traditional solution-phase synthesis, while the activated *N*-terminal glycopeptide **13** was prepared by solution-phase synthesis with solid-phase workup as described. The coupling between **12** and **13** was achieved in NMP and the product was readily isolated by precipitation. Target glycopeptide **11** was obtained in 89% overall yield (*28*).

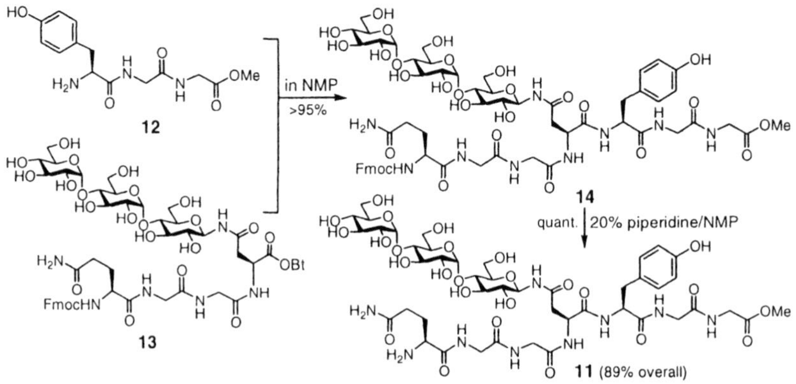

Scheme 3. Synthesis of glycopeptide 11 by fragment coupling

The method of solution-phase synthesis with solid-phase workup was also successfully applied to the synthesis of multivalent glycopeptides **15** and **16** that have 2 and 4 vicinal sugar chains respectively (*28*). To explain why the coupling of two glycosyl amino acids was especially efficient, we suggested that hydrogen bonds between the free carbohydrates might play some role. These glycopeptides should be difficult to prepare by traditional solid-phase synthesis.

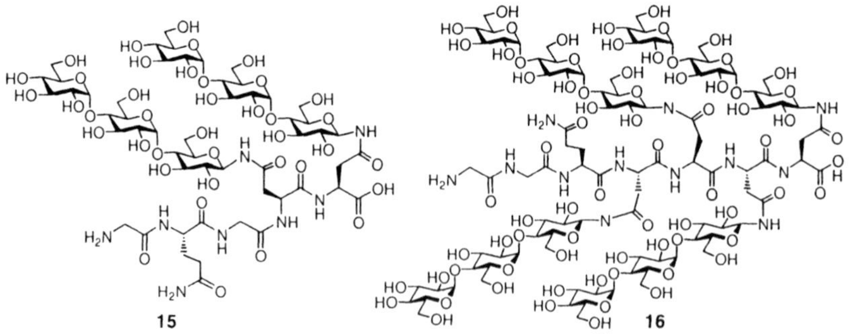

Recently, CD52 glycopeptide **17** with an acid-labile fucosidic linkage was also prepared by this new method (Scheme 4). CD52 is a glycosylphosphatidylinositol (GPI)-anchored glycopeptide that plays an important role in the human immune system (*29,30*) and human reproduction process (*31-33*).

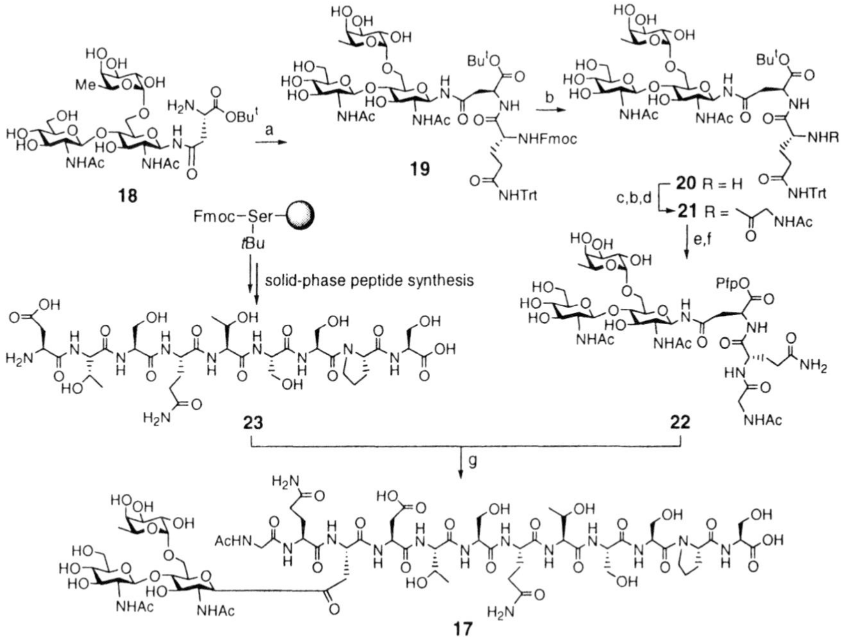

Scheme 4. *Synthesis of a CD52 glycopeptide by the new method. Reagents and conditions: (a) Fmoc-Gln(Trt)-OPfp, NMP, 96%; (b) 20% piperidine in NMP, 98%; (c) Fmoc-Gly-OPfp, NMP, 98%; (d) Ac_2O/MeOH, 91%; (e) 20% TFA in DCM; (f) DCC, PfpOH, NMP; (g) NMP, 42% (3 steps overall after HPLC).*

The *N*-terminal glycopeptide **21** was assembled from unprotected glycosyl asparagine **18** by solution-phase synthesis with solid-phase workup and utilizing the pentafluorophenyl (Pfp) ester of amino acids for the coupling reactions (*27*). The *C*-terminus of **21** was then deprotected with 20% TFA, and the product was treated with DCC and pentafluorophenol in NMP. The resultant active ester **22** was precipitated and washed by diethyl ether to form a product essentially free of DCC and DCU. As such, the new strategy enabled brief purification of the active ester without column chromatography. On the other hand, a nonapeptide segment **23** was synthesized by conventional solid-phase synthesis employing

Wang resin and Fmoc chemistry. The coupling reaction between **22** and **24** (3 eq.) was achieved in NMP to give the target structure **17** that was precipitated by diethyl ether and finally purified by HPLC (42%).

In summary, a new synthetic strategy for glycopeptides using unprotected glycosyl amino acids as building blocks and phase tags is described. It proved to be very efficient for a number of glycopeptides, including very complex ones that contain several vicinal sugar chains or contain oligosaccharides with acid-labile glycosidic linkages. It is also anticipated that the new synthetic strategy should be applicable to *O*-linked glycopeptides, a topic that is currently under further investigations.

Experimental Procedures *(27)*

Coupling Reaction of Amino Acid. Fmoc-Gln(Trt)-OPfp (284 mg, 0.375 mmol) was added to glycosyl asparagine **18** (92 mg, 0.125 mmol) in NMP (2 mL). After the homogeneous solution was stirred at room temperature for 2 h, diethyl ether (20 mL) was added with gentle stirring to yield a white precipitate. The mixture was centrifugated, and the supernatant was removed with a pipette carefully. The remaining precipitate was washed with diethyl ether (5 mL) twice and dried under a stream of nitrogen to give glycopeptide **19** (160 mg, 96%) as a white solid.

Peptide *N*-Terminal Deprotection. After a solution of crude **19** (160 mg, 0.12 mmol) dissolved in NMP (1.6 mL) and piperidine (0.4 mL) was stirred at room temperature for 1 h, diethyl ether (20 mL) was added to produce a white precipitate. After centrifugation, the supernatant was removed carefully, and the precipitate was washed with diethyl ether (5 mL) twice and dried under a stream of nitrogen to give glycopeptide **20** (130 mg, 98%) as an off-white solid.

Peptide *C*-Terminal Deprotection and Activation. Glycopeptide **21** was obtained by repeated *N*-terminal deprotection and coupling to amino acids. A part of **21** (5 mg, 0.005 mmol) was dissolved in 20% TFA in DCM (2 mL), and the solution was stirred at room temperature for 2 h. The reaction mixture was diluted with toluene (4 mL) and concentrated to dryness in vacuum. The residue was then dissolved in H_2O (5 mL). After washed with diethyl ether (5 mL) twice, the aqueous solution was lyophilized to afford a white solid. Upon its dissolution in NMP (1 mL), pentafluorophenol (9 mg, 0.05 mmol) and DCC (21 mg, 0.1 mmol) were added, and this homogeneous solution was stirred at room temperature for 5 h. The reaction mixture was diluted with diethyl ether (10 mL), and the resultant precipitate was isolated and washed as described above to give **22** as a wet white solid.

Coupling between Peptide and Glycopeptide. After the above crude **22** was dissolved in NMP (1 mL), to the solution was added free nonapeptide **23**

(10 mg, 0.011 mmol). The reaction mixture was stirred at room temperature for 12 h. Diethyl ether was added to yield a white precipitate that was isolated and washed as above. This solid product was further purified by HPLC using a C18 column (1.5 cm x 25 cm) with 2% *i*-PrOH in water containing 0.1% TFA as eluent (2 mL/min). The synthetic target **17** (R_t = 25.3 min) was obtained in 42% yield (3.8 mg from **21**, 0.0021 mmol) with a part of the starting materials **23** (5 mg, 0.005 mmol) and **21** (1.9 mg, 0.0015 mmol) recovered.

Acknowledgements

I acknowledge the great contributions of my coworkers who have appeared in the publications cited herein. Their hard work and great devotions made this research work possible. This project is supported by grants from the Research Corporation (Research Innovation Award RI 0663) and the Petroleum Research Funds of American Chemical Society (ACS PRF grant 37743-G). We are also grateful to Dr. S. Chen and Mr. H. Faulk for the MS measurements.

References

1. Varki, A. *Glycobiology* **1993**, *3,* 97.
2. Dwek, R. A. *Chem. Rev.* **1996**, *96,* 683.
3. Large, D. G.; Warren, C. D. *Glycopeptides and Related Compounds;* Marcel Dekker, Inc.: N.Y., 1997.
4. Herzner, H.; Reipen, T.; Schultz, M.; Kunz, H. *Chem. Rev.* **2000**, *100,* 4495.
5. Seitz, O. *ChemBioChem* **2000**, *1,* 214.
6. Kunz, H. *Pure Appl. Chem.* **1993**, *65,* 1223.
7. Meldal, M.; Bock, K. *Glycoconjugate J.* **1994**, *11,* 59.
8. Paulsen, H. *Angew. Chem. Int. Ed. Engl.* **1990**, *29,* 823.
9. Witte, K.; Seitz, O.; Wong, C.-H. *J. Am. Chem. Soc.* **1998**, *120,* 1979.
10. Wong, C.-H.; Schuster, M.; Wang, P.; Sears, P. *J. Am. Chem. Soc.* **1993**, *115,* 5893.
11. Mizuno, M.; Haneda, K.; Iguchi, R.; Muramoto, M.; T., K.; Aimoto, S.; Yamamoto, K.; Inazu, T. *J. Am. Chem. Soc.* **1999**, *121,* 284.
12. Deras, I. L.; Takegawa, K.; Kondo, A.; Kato, I.; Lee, Y. C. *Bioorg. Med. Chem. Lett.* **1998**, *8,* 1763.
13. Schuster, M.; Wang, P.; Paulson, J. C.; Wong, C.-H. *J. Am. Chem. Soc.* **1994**, *116,* 1135.
14. Witte, K.; Sears, P.; Martin, R.; Wong, C.-H. *J. Am. Chem. Soc.* **1997**, *119,* 2114.

15. Unverzagt, C. *Angew. Chem., Intl. Ed. Eng.* **1996**, *35,* 2350.
16. Roberge, J. Y.; Beebe, X.; Danishefsky, S. J. *J. Am. Chem. Soc.* **1998**, *120,* 3915.
17. Cohen-Anifeld, S.-T.; Lansbury, P. T. J. *J. Am. Chem. Soc.* **1993**, *115,* 10531.
18. Marcaurelle, L. A.; Mizoue, L. S.; Wilken, J.; Oldham, L.; Kent, S. B. H.; Handel, T. M.; Bertozzi, C. R. *Chem. Euro. J.* **2001**, *7,* 1129.
19. Marcaurelle, L. A.; Bertozzi, C. R. *J. Am. Chem. Soc.* **2001**, *123,* 1587.
20. Otvos, L. J.; Urge, L.; Hollosi, M.; Wroblewski, K.; Graczyk, G.; Fasman, G. D.; Thurin, J. *Tetrahedron Lett.* **1990**, *41,* 5889.
21. Reimer, K. B.; Meldal, M.; Kusumoto, S.; Fukase, K.; Bock, K. *J. Chem. Soc., Perkin Trans. 1* **1993**, 925.
22. Laczko, I.; Hollosi, M.; Urge, L.; Ugen, K. E.; Weiner, D. B.; Mantsch, H. H.; Thurin, J.; Otvos Jr, L. *Biochemistry* **1992**, *31,* 4282.
23. Osborn, H. M. I.; Khan, T. H. *Tetrahedron Lett.* **1999**, *55,* 1807.
24. Kunz, H.; Unverzagt, C. *Angew. Chem. Int. Ed. Engl.* **1988**, *27,* 1697.
25. Guo, Z.; Nakahara, Y.; Nakahara, Y.; Ogawa, T. *Carbohydr. Res.* **1997**, *303,* 373.
26. Wen, S.; Guo, Z. *Org. Lett.* **2001**, *3,* 3773.
27. Shao, N.; Xue, J.; Guo, Z. *J. Org. Chem.* **2003**, *68,* 9003.
28. Xue, J.; Guo, Z. *J. Org. Chem.* **2003**, *68,* 2713.
29. Treumann, A.; Lifely, M. R.; Schneider, P.; Ferguson, M. A. J. *J. Biol. Chem.* **1995**, *270,* 6088.
30. Hale, G.; Xia, M.-Q.; Tighe, H. P.; Dyer, J. S.; Waldmann, H. *Tissue Antigens* **1990**, *35,* 118.
31. Schroter, S.; Derr, P.; Conradt, H. S.; Nimtz, M.; Hale, G.; Kirchhoff, C. *J. Biol. Chem.* **1999**, *274,* 29862.
32. Diekman, A.; Norton, E. J.; Klotz, K. L.; Westbrook, V. A.; Shibahara, H.; Naaby-Hansen, S.; Flickinger, C. J.; Herr, J. C. *FESEB J.* **1999**, *13,* 1303.
33. Eccleston, E. D.; White, T. W.; Howard, J. B.; Hamilton, D. W. *Mol. Reprod. Develop.* **1994**, *37,* 110.

Indexes

Author Index

Subject Index

T

Z